LOGARITHMS and EXPONENTIALS

e
$\ln e$
$\log_2 8$

Essential Skills Practice

Workbook with

Answers

Chris McMullen, Ph.D.

Logarithms and Exponentials Essential Skills Practice Workbook with Answers
Chris McMullen, Ph.D.

Copyright © 2020, 2023 Chris McMullen, Ph.D.

www.improveyourmathfluency.com
www.monkeyphysicsblog.wordpress.com
www.chrismcmullen.com

All rights are reserved. However, educators or parents who purchase one copy of this workbook (or who borrow one physical copy from a library) may make and distribute photocopies of selected pages for instructional (non-commercial) purposes for their own students or children only.

Zishka Publishing
ISBN: 978-1-941691-32-8

Mathematics > Precalculus
Mathematics > Logarithms

Contents

Introduction	iv
1 Review of Exponents	5
2 Logarithm Basics	21
3 Logarithm Rules	31
4 Change of Base	48
5 Exponentials	56
6 Hyperbolic Functions	65
7 Graphs	83
8 Applications	98
9 Calculus	111
10 Complex Numbers	133
Answer Key	156

Introduction

The goal of this workbook is to help students master essential logarithm skills through explanations, examples, and practice.

- The first chapter reviews a variety of rules regarding the algebra of powers and roots. Since the definition of a logarithm is based on exponents, students who are not already fluent with this should begin with the first chapter.
- The second chapter discusses what a logarithm is and provides practice with a variety of problems that can be solved without using a calculator. Students who become fluent in these exercises should have a better understanding of what a logarithm is.
- Chapter 3 focuses on algebra rules for logarithms such as product and quotient rules. Chapter 4 introduces the change of base formula, which can be especially handy.
- Chapter 5 discusses exponentials and Chapter 6 covers hyperbolic functions.
- Graphs of logarithms, exponentials, and hyperbolic functions are the topic of Chapter 7.
- A variety of applications are included in Chapter 8.
- Chapter 10 presents complex numbers in a way that is accessible to students even if they have never seen complex numbers before.
- Chapter 9 covers the calculus of logarithms and exponentials. Students who have not studied calculus should skip Chapter 9.
- Answer key. Practice makes permanent, but not necessarily perfect. Check the answers at the back of the book and strive to learn from any mistakes. This will help to ensure that practice makes perfect.

1 Review of Exponents

1.1 Powers

An **exponent** (also called a **power**) appears above and to the right of a quantity. For example, in 4^3 the number 3 is the exponent and the number 4 is called the **base**. An exponent indicates repeated multiplication. For example, 4^3 is shorthand for $4 \times 4 \times 4$ and thus $4^3 = 64$. When the exponent is a positive integer, the exponent indicates how many times the base multiplies itself. For example, 2^6 has six 2's multiplying:
$$2^6 = 2 \times 2 \times 2 \times 2 \times 2 \times 2 = 64$$
An exponent of 2 is called a **square**. For example, 7^2 is read as "7 squared" and equals $7^2 = 7 \times 7 = 49$. An exponent of 3 is called a **cube**. For example, 5^3 is read as "5 cubed" and equals $5^3 = 5 \times 5 \times 5 = 125$.

When an exponent equals 1, the answer equals the base. For example, $12^1 = 12$. Any number raised to a power of 1 equals itself.

When an exponent equals 0 and the base is nonzero, the answer equals one regardless of what the value of the base is. For example, $6^0 = 1$. The reason for this is provided in Sec. 1.5.

A negative number raised to an odd power is negative, whereas a negative number raised to an even power is positive. For example, $(-2)^3 = (-2)(-2)(-2) = 4(-2) = -8$ is negative, whereas $(-2)^4 = (-2)(-2)(-2)(-2) = 4(4) = 16$ is positive.

Example 1. $12^2 = 12 \times 12 = 144$

Example 2. $3^4 = 3 \times 3 \times 3 \times 3 = 81$

Example 3. $4^1 = 4$

Example 4. $7^0 = 1$

Example 5. $(-4)^3 = -64$

Example 6. $(-10)^4 = 10{,}000$

1 Review of Exponents

Exercise Set 1.1

Directions: Answer each exponent problem without using a calculator.

1) $6^2 =$

2) $8^3 =$

3) $9^0 =$

4) $7^1 =$

5) $4^4 =$

6) $2^8 =$

7) $(-1)^{25} =$

8) $15^2 =$

9) $2^{10} =$

10) $6^4 =$

11) $(-3)^3 =$

12) $(-9)^2 =$

13) $(-2)^5 =$

14) $(-5)^4 =$

15) $18^1 =$

16) $50^3 =$

17) $(-30)^2 =$

18) $100^0 =$

19) $6^3 =$

20) $0^9 =$

21) $8^2 =$

22) $(-10)^3 =$

23) $(-20)^4 =$

24) $3^5 =$

1.2 Real Roots

The **radical** sign ($\sqrt[n]{}$) indicates a **root**, which is the opposite of a power in the following sense. It asks, which number raised to the nth power equals the value under the radical? For example, $\sqrt[4]{81}$ asks which number raised to the 4th power equals 81? One answer is 3 because $3^4 = 3 \times 3 \times 3 \times 3 = 9 \times 9 = 81$. When n is an even number like 4, there are two real answers; one is positive and the other is negative. For example, there are two answers to $\sqrt[4]{81}$. One answer is 3 and the other answer is -3. The reason that -3 is also an answer is because the minus signs cancel when the power is even. In this case, $(-3)^4 = (-3) \times (-3) \times (-3) \times (-3) = 9 \times 9 = 81$. We can combine these two solutions together concisely by writing $\sqrt[4]{81} = \pm 3$. In contrast, when n is odd, there is only one solution, which has the same sign as the value under the radical. For example, $\sqrt[3]{8} = 2$ since $2^3 = 2 \times 2 \times 2 = 8$ and $\sqrt[3]{-8} = -2$ since $(-2)^3 = (-2) \times (-2) \times (-2) = 4 \times (-2) = -8$.

When $n = 2$, the radical is called a **square root**. For a square root, it is customary not to write the 2. For example, $\sqrt{16}$ means the same as $\sqrt[2]{16}$ and equals ± 4 since $(\pm 4)^2 = 4 \times 4 = 16$. When $n = 3$, the radical is called a **cube root**. An example of a cube root is $\sqrt[3]{1000} = 10$ since $10^3 = 10 \times 10 \times 10 = 1000$.

Remember these rules regarding the signs:
- For an even root like $\sqrt{9}$, $\sqrt[4]{625}$, or $\sqrt[6]{1}$, use a $\pm$ sign to indicate that there are two possible answers. For example, $\sqrt[6]{1} = \pm 1$. (These are even roots because the value of n in $\sqrt[n]{}$ is an even number.)
- For an odd root like $\sqrt[3]{64}$ or $\sqrt[5]{-32}$, there is only one real answer and it has the same sign as the value under the radical. For example, compare $\sqrt[5]{32} = 2$ with $\sqrt[5]{-32} = -2$.

If you are interested in non-real roots, see Chapter 10.

1 Review of Exponents

Example 1. $\sqrt{121} = \pm 11$ since $(\pm 11)^2 = 121$

Example 2. $\sqrt[3]{27} = 3$ since $3^3 = 3 \times 3 \times 3 = 27$

Example 3. $\sqrt[3]{-27} = -3$ since $(-3)^3 = (-3) \times (-3) \times (-3) = -27$

Example 4. $\sqrt[4]{16} = \pm 2$ since $(\pm 2)^4 = 2 \times 2 \times 2 \times 2 = 16$

Example 5. $\sqrt[5]{100,000} = 10$ since $10^5 = 10 \times 10 \times 10 \times 10 \times 10 = 100,000$

Exercise Set 1.2

Directions: Determine all of the real answers without using a calculator.

1) $\sqrt{49} =$

2) $\sqrt[3]{216} =$

3) $\sqrt[4]{256} =$

4) $\sqrt[3]{-125} =$

5) $\sqrt{81} =$

6) $\sqrt[3]{512} =$

7) $\sqrt[4]{625} =$

8) $\sqrt[3]{-8000} =$

9) $\sqrt[9]{512} =$

10) $\sqrt[6]{1,000,000,000,000} =$

1.3 Negative Exponents

If a number is raised to a negative power, this is equivalent to raising the **reciprocal** of the number to the absolute value of the power. As an example, $\left(\frac{3}{4}\right)^{-2}$ is equivalent to $\left(\frac{4}{3}\right)^2$. The reason for this is given in Sec. 1.5.

$$\left(\frac{3}{4}\right)^{-2} = \left(\frac{4}{3}\right)^2 = \frac{4^2}{3^2} = \frac{16}{9}$$

Recall that the reciprocal of a fraction is found by swapping the roles of the numerator and denominator. For example, the reciprocal of $\frac{3}{4}$ is equal to $\frac{4}{3}$. To find the reciprocal of a whole number, divide one by the number. For example, the reciprocal of 7 is $\frac{1}{7}$.

Example 1. $25^{-1} = \left(\frac{1}{25}\right)^1 = \frac{1^1}{25^1} = \frac{1}{25}$

Example 2. $6^{-2} = \left(\frac{1}{6}\right)^2 = \frac{1^2}{6^2} = \frac{1 \times 1}{6 \times 6} = \frac{1}{36}$

Example 3. $\left(\frac{2}{3}\right)^{-4} = \left(\frac{3}{2}\right)^4 = \frac{3^4}{2^4} = \frac{3 \times 3 \times 3 \times 3}{2 \times 2 \times 2 \times 2} = \frac{81}{16}$

Example 4. $\left(\frac{7}{5}\right)^{-1} = \left(\frac{5}{7}\right)^1 = \frac{5^1}{7^1} = \frac{5}{7}$

Example 5. $(-2)^{-3} = \left(\frac{-1}{2}\right)^3 = \frac{(-1)^3}{2^3} = \frac{(-1) \times (-1) \times (-1)}{2 \times 2 \times 2} = -\frac{1}{8}$

Example 6. $(-3)^{-2} = \left(\frac{-1}{3}\right)^2 = \frac{(-1) \times (-1)}{3 \times 3} = \frac{1}{9}$

Example 7. $\left(-\frac{4}{5}\right)^{-1} = \left(\frac{-5}{4}\right)^1 = \frac{(-5)^1}{4^1} = -\frac{5}{4}$

1 Review of Exponents

Exercise Set 1.3

Directions: Determine all of the real answers without using a calculator.

1) $7^{-2} =$

2) $9^{-1} =$

3) $\left(\frac{13}{19}\right)^{-1} =$

4) $\left(\frac{5}{9}\right)^{-2} =$

5) $5^{-4} =$

6) $7^{-3} =$

7) $\left(\frac{1}{3}\right)^{-4} =$

8) $\left(\frac{4}{5}\right)^{-3} =$

9) $(-9)^{-2} =$

10) $(-1)^{-1} =$

11) $(-6)^{-3} =$

12) $(-5)^{-4} =$

13) $1^{-7} =$

14) $10^{-3} =$

15) $48^{-1} =$

16) $\left(\frac{5}{2}\right)^{-3} =$

17) $\left(-\frac{4}{9}\right)^{-2} =$

18) $\left(-\frac{1}{3}\right)^{-5} =$

19) $8^{-3} =$

20) $3^{-4} =$

21) $(-2)^{-9} =$

22) $\left(\frac{11}{12}\right)^{-1} =$

Logarithms and Exponentials Essential Skills Practice Workbook with Answers

1.4 Fractional Exponents

If a number is raised to a fractional exponent, the numerator has the effect of a whole number power while the denominator has the effect of a root. As an example, consider $8^{2/3}$, where the effect of the numerator (2) is the same as a square while the effect of the denominator (3) is the same as a cube root. These two effects combine together. Specifically, we may combine $8^2 = 64$ and $\sqrt[3]{64} = 4$ together as follows:
$$8^{2/3} = \sqrt[3]{8^2} = \sqrt[3]{64} = 4$$
Note that the order doesn't matter. If we find the cube root of 8 first and then find the square of this, we get the same answer:
$$8^{2/3} = \left(\sqrt[3]{8}\right)^2 = 2^2 = 4$$
The reason that the numerator has the effect of a power while the denominator has the effect of a root is given in Sec. 1.5.

For a fractional exponent that is negative, the minus sign has the same effect as in Sec. 1.3. First take the reciprocal of the base and then raise this to the absolute value of the exponent, like the example below.

$$\left(\frac{16}{81}\right)^{-3/4} = \left(\frac{81}{16}\right)^{3/4} = \left(\sqrt[4]{\frac{81}{16}}\right)^3 = \left(\pm\frac{3}{2}\right)^3 = \pm\frac{3^3}{2^3} = \pm\frac{27}{8}$$

Example 1. $9^{1/2} = \left(\sqrt{9}\right)^1 = (\pm 3)^1 = \pm 3$

Example 2. $32^{2/5} = \left(\sqrt[5]{32}\right)^2 = 2^2 = 4$

Example 3. $\left(\frac{27}{1000}\right)^{5/3} = \left(\sqrt[3]{\frac{27}{1000}}\right)^5 = \left(\frac{3}{10}\right)^5 = \frac{3^5}{10^5} = \frac{243}{100,000}$

Example 4. $\left(\frac{9}{16}\right)^{-3/2} = \left(\frac{16}{9}\right)^{3/2} = \left(\sqrt{\frac{16}{9}}\right)^3 = \left(\pm\frac{4}{3}\right)^3 = \pm\frac{4^3}{3^3} = \pm\frac{64}{27}$

Exercise Set 1.4

Directions: Determine all of the real answers without using a calculator.

1) $27^{4/3} =$

2) $256^{1/4} =$

3) $36^{1/2} =$

4) $64^{5/3} =$

5) $1{,}000{,}000^{5/6} =$

6) $32^{4/5} =$

7) $(-243)^{3/5} =$

8) $(-216)^{2/3} =$

9) $125^{-1/3} =$

10) $121^{-3/2} =$

11) $\left(\frac{4}{9}\right)^{5/2} =$

Logarithms and Exponentials Essential Skills Practice Workbook with Answers

12) $\left(\dfrac{625}{81}\right)^{3/4} =$

13) $\left(\dfrac{32}{243}\right)^{1/5} =$

14) $\left(\dfrac{1}{256}\right)^{-1/8} =$

15) $\left(-\dfrac{8}{27}\right)^{5/3} =$

16) $\left(-\dfrac{243}{1024}\right)^{2/5} =$

17) $160{,}000^{5/4} =$

18) $(-1)^{-8/7} =$

19) $225^{1/2} =$

20) $256^{-3/4} =$

21) $\left(-\dfrac{1}{100{,}000}\right)^{-9/5} =$

22) $\left(-\dfrac{343}{512}\right)^{-2/3} =$

1 Review of Exponents

1.5 Power Rules

An exponent that is a positive integer indicates repeated multiplication. For example, $x^3 = xxx$ has 3 x's multiplying one another. The number of x's multiplying in x^m is equal to m. Since the number of x's multiplying in $x^m x^n$ is equal to $m + n$, it follows that:
$$x^m x^n = x^{m+n}$$
It turns out that the formula above is true for all values of m and n, including fractions, zero, and negative values. If we let $m = 0$, this formula gives $x^0 x^n = x^n$. Provided that x is nonzero, we may divide both sides of the equation by x^n to get $x^0 = \frac{x^n}{x^n} = 1$. Any nonzero value raised to the power of zero is equal to one:
$$x^0 = 1 \text{ (if } x \neq 0\text{)}$$
Note that 0^0 is **indeterminate**. When x equals zero, we run into trouble if we attempt to divide both sides by x^n, since division by zero is a problem. (A nonzero value divided by zero is undefined, while zero divided by zero is indeterminate. Why? Consider the division problem $8 \div 4$, which asks, "Which number times 4 equals 8?" The answer is 2. If you try this with $1 \div 0$, this asks, "Which number times 0 equals 1?" This is undefined: you can't multiply 0 by a number and make 1. If you try this with $0 \div 0$, this asks, "Which number times 0 equals 0?" The answer is any real number. Any number times 0 equals zero. That's why $0 \div 0$ is indeterminate.)

In the case that $n = -m$, the formula gives $x^m x^{-m} = x^{m-m} = x^0$, which equals 1 if x is nonzero. This means that $x^m x^{-m} = 1$ if x is nonzero. Divide both sides by x^m to get:
$$x^{-m} = \frac{1}{x^m} \text{ (if } x \neq 0\text{)}$$
This shows that a negative exponent has the same effect as the absolute value of the exponent acting on the **reciprocal** of the base, which is exactly what we did in Sec. 1.3. The special case $m = 1$ shows that a power of minus one makes a reciprocal: $x^{-1} = \frac{1}{x}$ (if $x \neq 0$). A reciprocal is a **multiplicative inverse** in the sense that $xx^{-1} = x^{-1}x = 1$.

Consider the expression $(x^m)^n$. In this expression, the number of x^m's multiplying is equal to n. For example, if $m = 5$ and $n = 3$, we get $(x^5)^3 = x^5 x^5 x^5$. Since the number

of x^m's multiplying is equal to n and since the number of x's multiplying in each x^m is equal to m, all together there will be mn (meaning m times n) x's multiplying:
$$(x^m)^n = x^{mn}$$
For example, if $m = 5$ and $n = 3$, we get $(x^5)^3 = x^{15}$.

The expression $x^{1/n}$ represents the nth root of x:
$$x^{1/n} = \sqrt[n]{x}$$
One way to see this is as follows. Let the number of $x^{1/n}$'s multiplying one another be equal to n. Then the rule $x^m x^n = x^{m+n}$ will become $x^{1/n} x^{1/n} \cdots x^{1/n} = x^{1/n + 1/n + \cdots 1/n}$. Note that $\frac{1}{n} + \frac{1}{n} + \cdots + \frac{1}{n} = \frac{n}{n} = 1$ because there are n terms added together. The equation $x^{1/n} x^{1/n} \cdots x^{1/n} = x^1 = x$ states that when the number of $x^{1/n}$'s multiplied together equals n, the answer is x. This is exactly the definition of $\sqrt[n]{x}$, showing that $x^{1/n}$ is the same as $\sqrt[n]{x}$. For example, $\sqrt[3]{x}$ asks, "Which number cubed equals x?" This means that $\left(\sqrt[3]{x}\right)^3 = \sqrt[3]{x}\sqrt[3]{x}\sqrt[3]{x} = x$. Compare this to $x^{1/3} x^{1/3} x^{1/3} = x^{1/3+1/3+1/3} = x^1 = x$ to see that $x^{1/3}$ is equivalent to $\sqrt[3]{x}$. In general, $x^{1/n}$ is equivalent to $\sqrt[n]{x}$. As a special case of $x^{1/n} = \sqrt[n]{x}$, an exponent of one-half is equivalent to a square root:
$$x^{1/2} = \sqrt{x}$$

We may combine the formula $x^{1/n} = \sqrt[n]{x}$ with $(x^m)^n = x^{mn}$ to interpret a fractional power $x^{m/n}$ two different ways. One way shows that $x^{m/n}$ is equivalent to finding the nth root of x^m:
$$x^{m/n} = \sqrt[n]{x^m} = (x^m)^{1/n}$$
The other ways shows that $x^{m/n}$ is equivalent to raising the nth root of x to the mth power:
$$x^{m/n} = \left(\sqrt[n]{x}\right)^m = \left(x^{1/n}\right)^m$$
We explored these two ways in Sec. 1.4.

Consider the expression $(ax)^n$. The number of ax's multiplying together is n. That is, $(ax)^n = axaxax \cdots ax = (aaa \cdots a)(xxx \cdots x) = a^n x^n$, which shows that:
$$(ax)^n = a^n x^n$$

1 Review of Exponents

For example, $(2x)^3 = 2x2x2x = 2^3x^3 = 8x^3$. The special case $n = 1/2$ gives $(ax)^{1/2} = a^{1/2}x^{1/2}$. Since $x^{1/2} = \sqrt{x}$, this may be rewritten as:

$$\sqrt{ax} = \sqrt{a}\sqrt{x}$$

This allows us to factor out perfect squares. For example, $\sqrt{45} = \sqrt{9}\sqrt{5} = 3\sqrt{5}$. It can similarly be shown that:

$$\sqrt{\frac{x}{a}} = \frac{\sqrt{x}}{\sqrt{a}}$$

Example 1. $x^6 x^3 = x^{6+3} = x^9$

Example 2. $x^7 x^{-2} = x^{7+(-2)} = x^5$

Example 3. $\frac{x^9}{x^7} = x^{9-7} = x^2$

Example 4. $\frac{x^3}{x^{-2}} = x^{3-(-2)} = x^{3+2} = x^5$

Example 5. $x^4 x^{-4} = \frac{x^4}{x^4} = x^{4-4} = x^0 = 1$

Example 6. $(x^2)^6 = x^{2(6)} = x^{12}$

Example 7. $(3x)^4 = 3^4 x^4 = 81x^4$

Example 8. $\sqrt{9x} = \pm 3\sqrt{x}$

Example 9. $\sqrt{50} = \sqrt{25}\sqrt{2} = 5\sqrt{2}$

Example 10. $\sqrt{\frac{4}{9}} = \frac{\sqrt{4}}{\sqrt{9}} = \pm\frac{2}{3}$

Example 11. $\sqrt{x^8} = (x^8)^{1/2} = \pm x^4$

Example 12. $\sqrt[3]{x^6} = (x^6)^{1/3} = x^2$

Exercise Set 1.5

Directions: Simplify each expression, assuming $x \neq 0$.

1) $x^4 x^2 =$

2) $x^8 x^{-4} =$

3) $x^5 x^4 x =$

4) $x^2 x^{-3} =$

Logarithms and Exponentials Essential Skills Practice Workbook with Answers

5) $\dfrac{x^8}{x^3} =$

6) $\dfrac{x^2}{x^9} =$

7) $\dfrac{x^6}{x^6} =$

8) $\dfrac{x^6}{x^{-6}} =$

9) $\dfrac{x^5}{x^{-3}} =$

10) $\dfrac{x^{-3}}{x^{-4}} =$

11) $\dfrac{x^{-7}}{x} =$

12) $(x^4)^5 =$

13) $(-x^6)^3 =$

14) $(-x^3)^8 =$

15) $(x^5)^{-2} =$

16) $(x^{1/3})^6 =$

17) $(4x^2)^3 =$

18) $(3x^4)^{-5} =$

19) $\sqrt{36x^2} =$

20) $\sqrt{108x^3} =$

21) $\sqrt{\dfrac{25x^{12}}{49}} =$

22) $\sqrt{\dfrac{12x^8}{27}} =$

1.6 Euler's Number

Euler's number is a constant that is approximately equal to 2.71828. As we'll learn in Chapters 2 and onward, Euler's number is particularly important to logarithms and exponentials. Since we haven't yet learned about logarithms or exponentials, in this section we will explore how Euler's number relates to compound interest (which is a formula that involves an exponent).

The formula for **compound interest** is:

$$A = P\left(1 + \frac{r}{n}\right)^{nt}$$

- P represents the initial balance (it is called the **principal**).
- A represents the final balance (accounting for the principal plus the interest).
- r represents the **interest rate** (in decimal form; for example, 0.3 means 30%) for a unit time period (for example, an annual interest rate is per year).
- t represents the number of time periods that have elapsed thus far (or if you are making a prediction, it is the number of time periods that will have elapsed); note that t and r are both expressed in terms of the same time period (for an annual interest rate, t would be the number of years and r would be per year).
- n represents the number of times that interest is applied per time period (it is called the **compounding frequency**).

For example, suppose that $600 is invested in a savings account that pays interest at an annual rate of 3% compounded monthly, and we wish to predict what the balance would be in 10 years (without any other transactions or fees). In this case:
- $P = \$600$ is the initial balance
- $r = 0.03$ is the annual interest rate in decimal form (divide 3% by 100%)
- $t = 10$ years (for an annual interest rate, the unit time is one year)
- $n = 12$ (the interest is compounded 12 times per year, meaning once per month)

In this example, after 10 years the final balance will be (found using a calculator):

$$A = P\left(1 + \frac{r}{n}\right)^{nt} = \$600\left(1 + \frac{0.03}{12}\right)^{12(10)} = \$600(1.0025)^{120} = \$809.61$$

Logarithms and Exponentials Essential Skills Practice Workbook with Answers

How does the compound interest formula relate to Euler's number? We'll see. Let $P = 1$, $r = 1$ (corresponding to 100%), and $t = 1$ (corresponding to one unit time period). In this case, the formula becomes:

$$A = \left(1 + \frac{1}{n}\right)^n$$

In the limit that n (the compounding frequency) becomes infinitely large, the value of A (the final balance) approaches e (Euler's number).

$$e = \lim_{n \to \infty} \left(1 + \frac{1}{n}\right)^n$$

We encourage you to explore this numerically in the exercises that follow.

Example 1. Use a calculator to determine $\left(1 + \frac{1}{n}\right)^n$ for $n = 5$.

$$\left(1 + \frac{1}{5}\right)^5 = (1 + 0.2)^5 = (1.2)^5 = 2.48832$$

Exercise Set 1.6

Directions: Use a calculator to determine each of the following.

1) $\left(1 + \frac{1}{n}\right)^n$ for $n = 10$

2) $\left(1 + \frac{1}{n}\right)^n$ for $n = 100$

3) $\left(1 + \frac{1}{n}\right)^n$ for $n = 1000$

4) $\left(1 + \frac{1}{n}\right)^n$ for $n = 1{,}000{,}000$

1 Review of Exponents

An exclamation mark (!) represents a factorial. For example, 4! reads "4 factorial." A factorial means to multiply successively smaller integers until reaching 1. For example, $4! = 4 \times 3 \times 2 \times 1 = 24$. Two special cases are worth noting: $1! = 1$ and $0! = 1$. Note that 0! equals 1 (it is **not** zero). If it seems odd to you that $0! = 1$, consider this. For any positive integer n, we can write $n! = n(n-1)!$ provided that $0! = 1$. For example, 7! equals 7 times 6! Since $0! = 1$, for $n = 1$ we get $1! = 1(0!) = 1(1) = 1$.

Example 2. Use a calculator to determine $\frac{1}{0!} + \frac{1}{1!}$.

$$\frac{1}{0!} + \frac{1}{1!} = \frac{1}{1} + \frac{1}{1} = 1 + 1 = 2$$

Example 3. Use a calculator to determine $\frac{1}{0!} + \frac{1}{1!} + \frac{1}{2!}$.

$$\frac{1}{0!} + \frac{1}{1!} + \frac{1}{2!} = \frac{1}{1} + \frac{1}{1} + \frac{1}{2(1)} = 1 + 1 + \frac{1}{2} = 2.5$$

Directions: Use a calculator to determine each of the following.

5) $\frac{1}{0!} + \frac{1}{1!} + \frac{1}{2!} + \frac{1}{3!} =$

6) $\frac{1}{0!} + \frac{1}{1!} + \frac{1}{2!} + \frac{1}{3!} + \frac{1}{4!} =$

7) $\frac{1}{0!} + \frac{1}{1!} + \frac{1}{2!} + \frac{1}{3!} + \frac{1}{4!} + \frac{1}{5!} =$

8) $\frac{1}{0!} + \frac{1}{1!} + \frac{1}{2!} + \frac{1}{3!} + \frac{1}{4!} + \frac{1}{5!} + \frac{1}{6!} =$

Question: What will happen if we continue this series indefinitely?

2 Logarithm Basics

2.1 What Is a Logarithm?

Consider the problem $4^y = 23$. This problem asks, what exponent of 4 makes 23? In this problem, we're solving for the exponent. It would be challenging to determine the value of y in this case without using a calculator that has a logarithm function. We know that $4^2 = 16$ and that $4^3 = 64$, which tells us that y must lie between 2 and 3. We also know that $4^{2.5} = 4^{5/2} = \left(\sqrt{4}\right)^5 = 2^5 = 32$, which tells us that y lies between 2 and 2.5. It would be difficult to narrow the answer down very precisely by hand. A logarithm provides a way to solve for an exponent. In this case, $y = \log_4 23 \approx 2.261780978$ gives us the answer to $4^y = 23$. You can check that this works by entering $4^{2.261780978}$ on a scientific calculator. (You may be wondering how to enter $\log_4 23$ on your calculator. We'll answer that question in Chapter 4. In this chapter, we'll begin with the basics.)

A problem of the general form $x = b^y$ is equivalent to a logarithm of the form:
$$y = \log_b x$$
We read this as, "y equals log base b of x."
- b is the base of the logarithm. For example, in $\log_3 81 = 4$, the base is $b = 3$.
- x is the argument of the logarithm. For example, in $\log_3 81 = 4$, the argument is $x = 81$.
- y is the exponent of b that equals x. For example, $\log_3 81$ asks the question, "Which exponent of 3 equals 81?" The answer is $y = 4$ because $3^4 = 81$.

The logarithm function $\log_b x$ means, "Which exponent of b is equal to x?" For example, $\log_2 32$ means, "Which exponent of 2 is equal to 32?" The answer is $\log_2 32 = 5$ because $2^5 = 32$. The problem $y = \log_2 32$ is equivalent to the problem $2^y = 32$. In each case, our goal is to determine which power of 2 is equal to 32. In each case, $y = 5$ solves the problem.

2 Logarithm Basics

Example 1. Explain what $\log_5 25$ means in words, rewrite it as an equation with an exponent, and determine the answer.
- $\log_5 25$ asks, "Which exponent of 5 equals 25?"
- The problem $y = \log_5 25$ is equivalent to $5^y = 25$.
- The answer is $\log_5 25 = 2$ because $5^2 = 25$.

Exercise Set 2.1

Directions: Explain what each logarithm means in words, rewrite it as an equation with an exponent, and determine the answer without using a calculator.

1) $\log_4 64 =$

2) $\log_2 256 =$

3) $\log_{10} 1{,}000{,}000 =$

4) $\log_5 625 =$

Logarithms and Exponentials Essential Skills Practice Workbook with Answers

2.2 Negative Answers to Logarithms

The answer to a logarithm may be negative. As an example, consider $\log_2 0.25$. As we learned in Sec. 2.1, $\log_2 0.25$ asks, "Which exponent of 2 equals 0.25?" This logarithm is equivalent to $2^y = 0.25$. The answer is $\log_2 0.25 = -2$ because $2^{-2} = \left(\frac{1}{2}\right)^2 = 0.5^2 = 0.25$. (Recall that we reviewed negative exponents in Sec. 1.3.)

Example 1. Explain what $\log_5 0.008$ means in words, rewrite it as an equation with an exponent, and determine the answer.
- $\log_5 0.008$ asks, "Which exponent of 5 equals 0.008?"
- The problem $y = \log_5 0.008$ is equivalent to $5^y = 0.008$.
- Since $0.008 = \frac{8}{1000} = \frac{8 \div 8}{1000 \div 8} = \frac{1}{125}$, this is equivalent to $5^y = \frac{1}{125}$.
- The answer is $\log_5 0.008 = -3$ because $5^{-3} = \left(\frac{1}{5}\right)^3 = \frac{1}{5^3} = \frac{1}{125} = 0.008$.

Exercise Set 2.2

Directions: Explain what each logarithm means in words, rewrite it as an equation with an exponent, and determine the answer without using a calculator.

1) $\log_{10} 0.00001 =$

2) $\log_{25} 0.04 =$

2 Logarithm Basics

2.3 Base-10 Logarithms

Base-10 logarithms are common because it is often convenient to work with base ten. For example, the metric system which is based on powers of 10 is commonly used in science and engineering. Two common applications of base-10 logarithms include the loudness of sounds measured in decibels and the half-life of radioactive decays. A base-10 logarithm uses 10 as the base. For example, $\log_{10} 1000$ is a base-10 logarithm. It asks, "Which exponent of 10 equals 1000?" The answer is 3 because $10^3 = 1000$. Base-10 logarithms are easy to work with when the argument is a power of 10:

- If the argument is a power of 10 greater than 1, the answer is positive. Simply count the number of zeros. For example, $\log_{10} 100,000 = 5$ since 100,000 has 5 zeros. Note that $10^5 = 100,000$.
- If the argument is 1, the answer is zero. For example, $\log_{10} 1 = 0$ since $10^0 = 1$.
- If the argument is a power of 10 less than 1, the answer is negative. Add one to the number of zeros between the decimal point and the 1. For example, $\log_{10} 0.0001 = -4$ since 0.0001 has 3 zeros between the decimal point and the 1 (and since $3 + 1 = 4$). Note that $10^{-4} = \frac{1}{10^4} = \frac{1}{10,000} = 0.0001$.

Example 1. $\log_{10} 10,000,000 = 7$ because $10^7 = 10,000,000$ (there are 7 zeros)

Example 2. $\log_{10} 0.001 = -3$ because $10^{-3} = \frac{1}{1000} = 0.001$ (there are 2 zeros between the decimal point and the 1, and $2 + 1 = 3$; it is negative because $0.001 < 1$)

Exercise Set 2.3

Directions: Answer each logarithm problem without using a calculator.

1) $\log_{10} 10,000 =$

2) $\log_{10} 1,000,000 =$

Logarithms and Exponentials Essential Skills Practice Workbook with Answers

3) $\log_{10} 100 =$

4) $\log_{10} 0.000001 =$

5) $\log_{10} 0.1 =$

6) $\log_{10} 100,000,000 =$

7) $\log_{10} 1 =$

8) $\log_{10} 0.01 =$

9) $\log_{10} 1,000,000,000 =$

10) $\log_{10} 0.0000001 =$

11) $\log_{10} 1,000,000,000,000 =$

12) $\log_{10} 10,000,000,000 =$

13) $\log_{10} 0.0000000001 =$

14) $\log_{10} 0.000000000001 =$

15) $\log_{10}(10^{38}) =$

16) $\log_{10}(10^{-75}) =$

2.4 Exploring Logarithms

This section includes a variety of logarithms that can be solved without a calculator.

Example 1. $\log_7 49 = 2$ because $7^2 = 49$

Example 2. $\log_6 216 = 3$ because $6^3 = 216$

Example 3. $\log_2 0.125 = -3$ because $2^{-3} = \frac{1}{2^3} = \frac{1}{8} = 0.125$

Example 4. $\log_4 8 = 1.5$ because $4^{1.5} = 4^{3/2} = \left(\sqrt{4}\right)^3 = 2^3 = 8$ (Recall Sec. 1.4)

Exercise Set 2.4

Directions: Answer each logarithm problem without using a calculator.

1) $\log_3 81 =$

2) $\log_4 1024 =$

3) $\log_5 125 =$

4) $\log_{100} 1{,}000{,}000{,}000{,}000 =$

5) $\log_6 \left(\frac{1}{36}\right) =$

6) $\log_2 0.0625 =$

7) $\log_7 1 =$

8) $\log_{100} 0.000001 =$

9) $\log_8 0.125 =$

10) $\log_5 0.008 =$

11) $\log_{20} 160{,}000 =$

12) $\log_2 2048 =$

13) $\log_3 729 =$

14) $\log_{25} 625 =$

15) $\log_{20} 0.000125 =$

16) $\log_{25} 0.0016 =$

17) $\log_4 0.015625 =$

18) $\log_3 \frac{1}{81} =$

2 Logarithm Basics

19) $\log_4 32 =$

20) $\log_{27} 81 =$

21) $\log_{100} 10 =$

22) $\log_{16} 512 =$

23) $\log_9 243 =$

24) $\log_8 4 =$

25) $\log_{25} 0.00032 =$

26) $\log_{1000} 0.01 =$

27) $\log_{27} \left(\frac{1}{9}\right) =$

28) $\log_{64} 0.25 =$

29) $\log_{81} \left(\frac{1}{243}\right) =$

30) $\log_{32} \left(\frac{1}{16}\right) =$

Logarithms and Exponentials Essential Skills Practice Workbook with Answers

2.5 Natural Logarithms

In many common applications of logarithms, such as the half-life of radioactive decays or a discharging capacitor, the base of the logarithm naturally turns out to be Euler's number, which is approximately $e \approx 2.71828$ (Sec. 1.6). We call this a **natural logarithm**. The natural logarithm is so common in math and science that we give it a special name. We use $\ln(y)$ to represent the natural logarithm, whereas we use $\log_b(y)$ for all other logarithms. Note that $\ln(y) = \log_e(y)$. The natural logarithm has e as its base.

Since the general logarithm $x = \log_b(y)$ is equivalent to $y = b^x$, the natural logarithm $x = \ln(y)$ is equivalent to $y = e^x$ (since a natural logarithm has base e).

Almost all natural logarithm calculations require using a calculator to determine the answer numerically. One exception is $\ln 1 = 0$ because $e^0 = 1$. Another exception is when the argument is expressed as a power of Euler's number, such as $\ln(e^3) = 3$ or $\ln(e^{7.5}) = 7.5$ (in these cases, the answer simply equals the exponent of e).

Look at your calculator. Most scientific calculators have a log button, a ln button, or both. Some scientific calculators have other ways of accessing the logarithm function, such as through a menu. (You may need to read your owner's manual, look for help on the calculator's website, or Google how to find a logarithm with the specific model of your calculator.) If a calculator has both log and ln buttons, the log button is probably a base-10 logarithm and ln is a natural logarithm. Some calculators have a log button with the ln feature accessible by first pressing the 2nd button. If a calculator only has a log button and there doesn't appear to be any ln option, there is an easy way to check which base the log button uses. Take the logarithm of 1000 using that button. If the answer is 3, it is a base-10 logarithm. If the answer is 6.907755..., it is a natural logarithm. (To make matters worse, a few authors of textbooks or articles use $\log x$ without any base to mean a natural logarithm, which can cause confusion.)

Once you know how to find a natural logarithm with your calculator, you are ready to attempt the problems of this section.

Example 1. $\ln 7 \approx 1.95$

Example 1. $\ln 0.95 \approx -0.0513$

Exercise Set 2.5

Directions: Use a calculator to determine each answer to 3 significant figures.

1) $\ln 5 \approx$

2) $\ln 2 \approx$

3) $\ln 10 \approx$

4) $\ln 0.5 \approx$

5) $\ln 1 \approx$

6) $\ln 0.999 \approx$

7) $\ln 25 \approx$

8) $\ln 100 \approx$

9) $\ln 0.367879 \approx$

10) $\ln 0.007 \approx$

11) $\ln 489 \approx$

12) $\ln 23{,}492 \approx$

13) $\ln 0.000001 \approx$

14) $\ln 10^{15} \approx$

15) $\ln\left(\frac{2}{3}\right) \approx$

16) $\ln \sqrt{2} \approx$

17) $\ln(49^3) \approx$

18) $\ln(8^{-3}) \approx$

19) $\ln(e^5) \approx$

20) $\ln\left(\frac{1}{e^8}\right) \approx$

3 Logarithm Rules

3.1 Cancellation Equations

In Chapter 2, we learned that the equation $y = \log_b x$ is equivalent to the equation $x = b^y$. If we substitute $x = b^y$ into $y = \log_b x$, we get the following equation, which is called a cancellation equation.

$$y = \log_b(b^y)$$

For example, $\log_{10} 10^3 = 3$, which agrees with $\log_{10} 1000 = 3$. If instead we substitute $y = \log_b x$ into $x = b^y$, we get a different cancellation equation.

$$x = b^{\log_b x}$$

For example, $10^{\log_{10} 1000} = 1000$, which agrees with $10^3 = 1000$. These cancellations work for any base (as long as the base is consistent within the equation), including a natural logarithm (Sec. 2.5), for which the base is Euler's number (e):

$$y = \ln(e^y)$$
$$x = e^{\ln x}$$

The natural logarithm ($\ln x$) and exponential function (e^x) are **inverse** functions. (We will explore exponential functions in Chapter 5.) The cancellation equations for natural logarithms follow from this inverse relationship. If you exponentiate a number and then apply the natural logarithm, the exponentiation and logarithm cancel out, which is why $\ln(e^y)$ equals y. Similarly, if you first take the natural logarithm of a number and then exponentiate that, the two functions cancel out, which is why $e^{\ln x}$ equals x.

Example 1. $\log_2(2^{5/3}) = \frac{5}{3}$ **Example 2.** $\ln e^{1.75} = 1.75$

Example 3. $5^{\log_5 47} = 47$ **Example 4.** $e^{\ln 6.3} = 6.3$

3 Logarithm Rules

Exercise Set 3.1

Directions: Answer each logarithm problem without using a calculator.

1) $\log_5(5^8) =$

2) $\log_{10}(10^{0.4}) =$

3) $\ln(e^9) =$

4) $\log_3(3^{-4/9}) =$

5) $6^{\log_6 3} =$

6) $42^{\log_{42}(2/3)} =$

7) $e^{\ln 2} =$

8) $10^{\log_{10} 4.17} =$

9) $\log_{10}(10^{123}) =$

10) $\ln(e^0) =$

11) $\ln\left(e^{\sqrt{2}}\right) =$

12) $\log_7(7^{1/7}) =$

13) $17^{\log_{17} 0.001} =$

14) $e^{\ln 0.8} =$

15) $e^{\ln 1} =$

16) $9^{\log_9 e} =$

17) $\log_2[2^{(3^4)}] =$

18) $\ln(e^{-1/\pi}) =$

19) $t^{\log_t \sqrt{3}} =$

20) $\log_x(x^{-5.2}) =$

Logarithms and Exponentials Essential Skills Practice Workbook with Answers

3.2 Sum of Logarithms

Let's define one variable (w) to equal the product of two other variables (p and q):
$$w = pq$$
We may apply the cancellation equation to each variable: $w = b^{\log_b w}$, $p = b^{\log_b p}$, and $q = b^{\log_b q}$. In each case, the logarithm and its inverse effectively cancel out, as we saw in Sec. 3.1. Substitute these cancellation equations into the previous equation.
$$b^{\log_b w} = b^{\log_b p} b^{\log_b q}$$
We may replace w with pq, since $w = pq$.
$$b^{\log_b(pq)} = b^{\log_b p} b^{\log_b q}$$
On the right-hand side, note that $b^{\log_b p} b^{\log_b q} = b^{\log_b p + \log_b q}$ according to the rule $x^m x^n = x^{m+n}$ (Sec. 1.5).
$$b^{\log_b(pq)} = b^{\log_b p + \log_b q}$$
This equation has the structure $b^t = b^u$, where $t = \log_b(pq)$ and $u = \log_b p + \log_b q$. In order for $b^t = b^u$ to be true, it must be true that $t = u$, which may be written as:
$$\log_b(pq) = \log_b p + \log_b q$$
This is called the **sum of logs** formula. It may also be written with the left and right sides swapped. (If $t = u$, it must also be true that $u = t$. Either way, both sides are equal.)
$$\log_b p + \log_b q = \log_b(pq)$$
If the base is Euler's number (e), this can be written with natural logarithms as:
$$\ln p + \ln q = \ln(pq)$$
We will apply these relations in the examples and exercises.

Example 1. $\log_3 0.5 + \log_3 18 = \log_3(0.5 \times 18) = \log_3 9 = 2$ since $3^2 = 9$

Example 2. $\ln 0.1 + \ln(10e) = \ln(0.1 \times 10e) = \ln e = 1$

Example 3. $\ln(4x) + \ln(3x - 2) = \ln[4x(3x - 2)] = \ln(12x^2 - 8x)$

Exercise Set 3.2

Directions: Answer each logarithm problem without using a calculator.

1) $\log_2(0.2) + \log_2 320 =$

2) $\log_{10} 40 + \log_{10} 25 =$

3) $\log_6 4 + \log_6 9 =$

4) $\ln(0.25e) + \ln(4e) =$

5) $\log_4 2 + \log_4 \sqrt{2} + \log_4 32 + \log_4 \sqrt{8} =$

Directions: Reduce each expression down to a single logarithm.

6) $\log_{10}(3x) + \log_{10}(4x) =$

7) $\log_2(x - 3) + \log_2(x + 3) =$

8) $\ln x + \ln(2x) + \ln(3x) =$

9) $\log_{10}|\tan x| + \log_{10}|\cos x| =$

10) $\ln \sqrt{7x} + \ln \sqrt{7x} =$

3.3 Difference of Logarithms

Let's define one variable (w) to equal the quotient of two other variables (p and q):

$$w = \frac{p}{q}$$

We will apply the cancellation equation to each variable, like we did in the previous section: $w = b^{\log_b w}$, $p = b^{\log_b p}$, and $q = b^{\log_b q}$. Substitute the cancellation equations into the previous equation.

$$b^{\log_b w} = \frac{b^{\log_b p}}{b^{\log_b q}}$$

We may replace w with $\frac{p}{q}$, since $w = \frac{p}{q}$.

$$b^{\log_b\left(\frac{p}{q}\right)} = \frac{b^{\log_b p}}{b^{\log_b q}}$$

On the right-hand side, note that $\frac{b^{\log_b p}}{b^{\log_b q}} = b^{\log_b p - \log_b q}$ according to the rule $\frac{x^m}{x^n} = x^{m-n}$ (Sec. 1.5).

$$b^{\log_b\left(\frac{p}{q}\right)} = b^{\log_b p - \log_b q}$$

This equation has the structure $b^t = b^u$, where $t = \log_b\left(\frac{p}{q}\right)$ and $u = \log_b p - \log_b q$. In order for $b^t = b^u$ to be true, it must be true that $t = u$, which may be written as:

$$\log_b\left(\frac{p}{q}\right) = \log_b p - \log_b q$$

This is called the **difference of logs** formula. It may also be written as:

$$\log_b p - \log_b q = \log_b\left(\frac{p}{q}\right)$$

If the base is Euler's number (e), this can be written with natural logarithms as:

$$\ln p - \ln q = \ln\left(\frac{p}{q}\right)$$

We will apply these relations in the examples and exercises.

Example 1. $\log_{12} 288 - \log_{12} 2 = \log_{12}\left(\frac{288}{2}\right) = \log_{12} 144 = 2$ since $12^2 = 144$

Example 2. $\ln(42x^7) - \ln(14x^3) = \ln\left(\frac{42x^7}{14x^3}\right) = \ln(3x^4)$

Exercise Set 3.3

Directions: Answer each logarithm problem without using a calculator.

1) $\log_{10} 2500 - \log_{10}(2.5) =$

2) $\log_3 486 - \log_3 6 =$

3) $\ln(7e) - \ln 7 =$

4) $\log_3 \sqrt{45} - \log_3 \sqrt{5} =$

5) $\log_6 1 - \log_6 4 - \log_6 9 =$

Directions: Reduce each expression down to a single logarithm.

6) $\log_5(24x^8) - \log_5(6x^3) =$

7) $\ln(72x) - \ln(4x) - \ln(3x) =$

8) $\log_{10}(x-5) - \log_{10}(x+2) - \log_{10}(x-2) =$

9) $\ln|\cos x| - \ln|\sin x| =$

10) $\log_2 \sqrt{48x} - \log_2 \sqrt{3x} =$

3.4 Coefficient of a Logarithm

Let's define one variable (w) to equal the exponent (a) of another variable (p):
$$w = p^a$$
We may apply the cancellation equation to each variable, like we did in the previous sections: $w = b^{\log_b w}$ and $p = b^{\log_b p}$. Substitute these cancellation equations into the previous equation.
$$b^{\log_b w} = \left(b^{\log_b p}\right)^a$$
We may replace w with p^a, since $w = p^a$.
$$b^{\log_b(p^a)} = \left(b^{\log_b p}\right)^a$$
On the right-hand side, note that $\left(b^{\log_b p}\right)^a = b^{a \log_b p}$ according to the rule $(x^m)^n = x^{mn}$ (Sec. 1.5).
$$b^{\log_b(p^a)} = b^{a \log_b p}$$
This equation has the structure $b^t = b^u$, where $t = \log_b(p^a)$ and $u = a \log_b p$. In order for $b^t = b^u$ to be true, it must be true that $t = u$, which may be written as:
$$\log_b(p^a) = a \log_b p$$
This formula may also be written with the left and right sides swapped.
$$a \log_b p = \log_b(p^a)$$
If the base is Euler's number (e), this can be written with natural logarithms as:
$$a \ln p = \ln(p^a)$$
Let's verify that this works for $\log_{10}(10^3)$. According to the formula, $\log_{10}(10^3) = 3 \log_{10} 10 = 3(1) = 3$, which agrees with $\log_{10}(10^3) = \log_{10}(1000) = 3$.

Example 1. $\log_{10}(100^{2.7}) = 2.7 \log_{10}(100) = 2.7(2) = 5.4$ since $10^2 = 100$

Example 2. $\ln(e^3) = 3 \ln e = 3(1) = 3$

Example 3. $\ln(x^2) = 2 \ln x$

Exercise Set 3.4

Directions: Answer each logarithm problem without using a calculator.

1) $\log_2(8^{7.5}) =$

2) $\log_{10}(0.1^{3/2}) =$

3) $\log_6(36^{0.3}) =$

4) $\ln(e^\pi) =$

5) $\log_3 \sqrt{243} =$

Directions: Rewrite each expression without any powers or roots in the argument.

6) $\log_{10}(x^{5/3}) =$

7) $5 \ln(x^{4.2}) =$

8) $\log_2(6^5 x^5) =$

9) $\log_{10}(\sin^2 x) =$

10) $\ln \sqrt{5x} =$

3.5 Negative of a Logarithm

If we set $a = -1$ in the formula $\log_b(p^a) = a \log_b p$, we get:
$$\log_b(p^{-1}) = -\log_b p$$
Since $p^{-1} = \frac{1}{p}$ (Sec. 1.5), we may also write this as:
$$\log_b\left(\frac{1}{p}\right) = -\log_b p$$
As always, these formulas also apply to natural logs:
$$\ln(p^{-1}) = -\ln p$$
$$\ln\left(\frac{1}{p}\right) = -\ln p$$
The sign in front of any logarithm may be changed by changing the sign of the exponent of the argument:
$$\log_b(p^{-a}) = -\log_b(p^a)$$
Since $p^{-a} = \frac{1}{p^a}$ (Sec. 1.5), we may also write this as:
$$\log_b\left(\frac{1}{p^a}\right) = -\log_b(p^a)$$
In terms of natural logarithms, these formulas are:
$$\ln(p^{-a}) = -\ln(p^a)$$
$$\ln\left(\frac{1}{p^a}\right) = -\ln(p^a)$$
For example, compare $\log_2 8 = 3$ with $\log_2\left(\frac{1}{8}\right) = -3$.

Example 1. $\log_7(49^{-1}) = -\log_7(49) = -2$ since $7^2 = 49$

Example 2. $\ln\left(\frac{1}{e}\right) = \ln(e^{-1}) = -\ln e = -1(1) = -1$

Example 3. $\ln\left(\frac{1}{x}\right) = \ln(x^{-1}) = -\ln x$

3 Logarithm Rules

Exercise Set 3.5

Directions: Answer each logarithm problem without using a calculator.

1) $\log_{10}(10^{-1}) =$

2) $\log_9 \left(\frac{1}{81}\right) =$

3) $\log_6(6^{-2}) =$

4) $\ln(e^{-3}) =$

5) $\log_4 \left(\frac{1}{64}\right) =$

Directions: Rewrite each expression without any fractions or exponents in the argument.

6) $\log_2(3^{-1}x^{-1}) =$

7) $-6\ln\left(\frac{1}{2x}\right) =$

8) $\log_{10}\left(\frac{1}{3xy}\right) =$

9) $\log_{10}\left|\frac{1}{\cos x}\right| =$

10) $\ln\left(\frac{1}{4x} + \frac{3}{4x}\right) =$

3.6 Combining Rules

The problems in this section combine two or more rules from this chapter. Following is a list of logarithm rules. The cancellation equations are:
$$y = \log_b(b^y)$$
$$x = b^{\log_b x}$$
$$y = \ln(e^y)$$
$$x = e^{\ln x}$$

The sum of logarithms formulas are:
$$\log_b(pq) = \log_b p + \log_b q$$
$$\ln(pq) = \ln p + \ln q$$

The difference of logarithms formulas are:
$$\log_b\left(\frac{p}{q}\right) = \log_b p - \log_b q$$
$$\ln\left(\frac{p}{q}\right) = \ln p - \ln q$$

The rules for an exponent of the argument are:
$$\log_b(p^a) = a \log_b p$$
$$\ln(p^a) = a \ln p$$

The rules for the reciprocal of the argument are:
$$\log_b\left(\frac{1}{p}\right) = \log_b(p^{-1}) = -\log_b p$$
$$\log_b\left(\frac{1}{p^a}\right) = \log_b(p^{-a}) = -\log_b(p^a) = -a \log_b p$$
$$\ln\left(\frac{1}{p}\right) = \ln(p^{-1}) = -\ln p$$
$$\ln\left(\frac{1}{p^a}\right) = \ln(p^{-a}) = -\ln(p^a) = -a \ln p$$

Example 1. $2 \log_5 10 - \log_5 4 = \log_5(10^2) - \log_5 4 = \log_5\left(\frac{10^2}{4}\right) = \log_5 25 = 2$

Example 2. $3 \ln x + \ln(x^2) = \ln(x^3) + \ln(x^2) = \ln[(x^3)(x^2)] = \ln x^5$ or $5 \ln x$

Exercise Set 3.6

Directions: Answer each logarithm problem without using a calculator.

1) $\log_6 54 + \log_6 12 - \log_6 18 =$

2) $2 \log_3 2 - 2 \log_3 6 =$

3) $8e^{-3 \ln 4} =$

4) $12 \log_4(2^3) =$

5) $5 \ln\left[\left(\frac{1}{e^5}\right)^8\right] =$

6) $\log_{10}\left(100^{6 \log_3 9}\right) =$

7) $36 \log_{100} \sqrt{\frac{1}{10}} =$

8) $\log_6\left(\frac{2}{3}\right) + \log_6\left(\frac{1}{24}\right) =$

9) $6 \ln\left[\ln\left(e^{e^{-5}}\right)\right] =$

10) $3 \log_2 \sqrt{8} + 5 \log_2 \sqrt{32} =$

Logarithms and Exponentials Essential Skills Practice Workbook with Answers

Directions: Reduce each expression down to a single logarithm without any fractions or exponents in the argument.

11) $4\log_7 x^2 - 3\log_7 x =$

12) $\log_5(x+5) + \log_5(x-5) - \log_5(x^2-25) =$

13) $5\ln\left(\frac{4}{x}\right) + 7\ln\left(\frac{x}{2}\right) + \ln x =$

14) $\log_{10}\left(\frac{x}{y}\right) - \log_{10}(4x) =$

15) $6\log_3 x^3 - 3\log_3 x^4 + 4\log_3 x^2 =$

16) $4x^6 e^{-5\ln x} =$

17) $\log_2(x^{1/2}) - \frac{1}{3}\log_2 x =$

18) $\ln(\sin^2 x) - \ln(\cos^2 x) =$

19) $3\log_{10}\left(\frac{x^3}{y^4}\right) - 6\log_{10}\left(\frac{x}{y^2}\right) =$

20) $\ln\sqrt{x^5} + \ln\sqrt{x^3} - 6\ln(x^2) =$

3.7 Solving Equations with Logarithms

Many equations involving logarithms and one variable can be solved as follows:
- If there is more than one term involving a logarithm, see if you can apply the sum or difference of logarithms formulas to reduce the number of terms that involve logarithms.
- Attempt to isolate the logarithm term. If you can get the logarithm term all by itself on one side of the equation, like $\log_b x = c$, then you can exponentiate both sides of the equation by using the same base as the logarithm. For example, $\log_b x = c$ becomes $b^{\log_b x} = b^c$. Apply the cancellation equation ($b^{\log_b x} = x$) to remove the logarithm: $x = b^c$. The examples that follow illustrate this idea.

Example 1. $2 \log_3 x + 4 = 14$
- Subtract 4 from both sides: $2 \log_3 x = 10$
- Divide by 2 on both sides: $\log_3 x = 5$ (The logarithm is now isolated.)
- Exponentiate both sides (with base 3): $3^{\log_3 x} = 3^5$
- Apply the cancellation equation: $x = 3^5$
- The answer is $x = 3^5 = 243$

Check the answer: Plug $x = 243$ into the original equation. Since $2 \log_3 243 + 4 = 2(5) + 4 = 10 + 4 = 14$, the answer checks out.

Example 2. $\ln(x^2) - \ln x = 1$
- Use the difference of logarithms formula: $\ln\left(\frac{x^2}{x}\right) = 1$
- Simplify the left side: $\ln x = 1$
- Exponentiate both sides (with base e): $e^{\ln x} = e^1$
- Apply the cancellation equation: $x = e^1$
- The answer is $x = e^1 = e$

Check the answer: Plug $x = e$ into the original equation. Since $\ln(e^2) - \ln e = 2 - 1 = 1$, the answer checks out.

Logarithms and Exponentials Essential Skills Practice Workbook with Answers

Exercise Set 3.7

Directions: Solve for the unknown in each equation without using a calculator.

1) $6 \log_5 x = 18$

2) $2 \log_9 x - 1 = 3$

3) $4 - \log_2 x = 9$

4) $\log_3 \left(\frac{x}{2}\right) = 4$

5) $\log_{10}(x^2) + 2 = 8$

6) $\log_6(x^5) - \log_6(x^2) = 6$

7) $\log_4 \sqrt{x} + \log_4 \sqrt{x} = 5$

8) $\log_9(6 - 3x) = 2$

9) $\log_3(x + 3) + \log_3(x - 3) = 3$

10) $\log_2(5x - 6) - \log_2(2x + 3) = 0$

Logarithms and Exponentials Essential Skills Practice Workbook with Answers

Directions: Solve for the unknown in each equation. You may use a calculator for these.

11) $4 \ln x = 12$

12) $\ln(6x^4) - \ln(2x^2) = 3$

13) $\ln x + \ln(3x + 2) = 1$

14) $\ln(\ln x) = 1$

15) $\ln|\sin x| = -1$ (Assume that the answer lies in Quadrant I.)

4 Change of Base

4.1 The Change of Base Formula

Consider the equation $y = \log_b x$, which corresponds to $x = b^y$ (Chapter 2). If we take the logarithm with respect to a different base, which we will call a, of both sides of $x = b^y$, we get $\log_a x = \log_a(b^y)$. From Sec. 3.4, we may rewrite the right side of the equation as follows: $\log_a x = y \log_a b$. If we plug $y = \log_b x$ into this equation, we get $\log_a x = \log_b x \log_a b$. If we divide both sides by $\log_a b$, we get the following equation: $\frac{\log_a x}{\log_a b} = \log_b x$. If we simply swap the sides, we may rewrite this equation as:

$$\log_b x = \frac{\log_a x}{\log_a b}$$

This is called the **change of base** formula. If you need to take a logarithm with respect to base b, you can express that logarithm in terms of logarithms with respect to base a using the change of base formula.

The change of base formula allows you to compute any logarithm on a calculator using the calculator's natural logarithm (base e) or common logarithm (base 10) functions. For example, to determine $\log_7 4.3$ on a calculator, the change of base formula allows you to enter $\frac{\ln 4.3}{\ln 7}$ (using natural logarithms) or $\frac{\log_{10} 4.3}{\log_{10} 7}$ (using base 10) instead of using base 7.

Setting $a = e$ lets you change the base of a logarithm into natural logarithms:

$$\log_b x = \frac{\ln x}{\ln b}$$

Setting $a = 10$ lets you change the base of a logarithm into base-10 logarithms:

$$\log_b x = \frac{\log_{10} x}{\log_{10} b}$$

We'll use this formula to do calculations by hand in Sec. 4.2 and with a calculator in Sec. 4.3. In this section, we'll focus on how to rewrite a logarithm in a new base.

Example 1. Rewrite $\log_3 12$ in base 10.

Plug $b = 3$, $x = 12$, and $a = 10$ into the change of base formula.

$$\log_3 12 = \frac{\log_a x}{\log_a b} = \frac{\log_{10} 12}{\log_{10} 3}$$

Note: We'll perform calculations with the change of base formula in Sec.'s 4.2-4.3. For now, the focus is on how to write the change of base formula for a specified base.

Example 2. Rewrite $\log_5 0.4$ in base e.

Plug $b = 5$ and $x = 0.4$ into the change of base formula, using natural logarithms for the new base.

$$\log_5 0.4 = \frac{\ln x}{\ln b} = \frac{\ln 0.4}{\ln 5}$$

Exercise Set 4.1

Directions: Rewrite each logarithm in the specified base. You don't need to perform the calculation. You just need to show how to rewrite it with the change of base formula.

1) Rewrite $\log_9 5$ in base 10.

2) Rewrite $\log_2 \left(\frac{1}{3}\right)$ in base e.

3) Rewrite $\log_5 \pi$ in base 2.

4.2 Changing the Base of a Logarithm

In this section, we will verify that the change of base formula works for a variety of calculations that can be performed without using a calculator.

Example 1. Rewrite $\log_2 64$ in base 4. Verify that the two sides of the formula are equal. Plug $b = 2$, $x = 64$, and $a = 4$ into the change of base formula.
$$\log_2 64 = \frac{\log_a x}{\log_a b} = \frac{\log_4 64}{\log_4 2}$$

- The left side equals $\log_2 64 = 6$ because $2^6 = 64$.
- The right side equals $\frac{\log_4 64}{\log_4 2} = \frac{3}{1/2} = 6$ because $4^3 = 64$ and $4^{1/2} = \sqrt{4} = 2$. Note: To divide by a fraction, multiply by its reciprocal: $\frac{3}{1/2} = 3 \div \frac{1}{2} = 3 \times \frac{2}{1} = \frac{6}{1} = 6$ or $\frac{3}{1/2} = \frac{3}{0.5} = \frac{3(2)}{0.5(2)} = \frac{6}{1} = 6$.
- Both sides equal 6, which verifies the formula for this example.

Example 2. Rewrite $\log_{1000} 1,000,000$ in base 10. Verify that the two sides of the formula are equal.
Plug $b = 1000$, $x = 1,000,000$, and $a = 10$ into the change of base formula.
$$\log_{1000} 1,000,000 = \frac{\log_a x}{\log_a b} = \frac{\log_{10} 1,000,000}{\log_{10} 1000}$$

- The left side equals $\log_{1000} 1,000,000 = 2$ because $1000^2 = 1,000,000$.
- The right side equals $\frac{\log_{10} 1,000,000}{\log_{10} 1000} = \frac{6}{3} = 2$ because $10^6 = 1,000,000$ and $10^3 = 1000$.
- Both sides equal 2, which verifies the formula for this example.

Exercise Set 4.2

Directions: Rewrite each logarithm in the specified base. Calculate each side of the formula to verify that both sides of the change of base formula are equal for each case.

1) $\log_3 81$ in base 9

2) $\log_8 512$ in base 2

3) $\log_{16} 256$ in base 4

4) $\log_{100} 0.000001$ in base 10

5) $\log_2 4096$ in base 16

6) $\log_9 27$ in base 3

4.3 Finding Logarithms with a Calculator

Suppose that you need to determine $\log_2 5$ on your calculator, but that your calculator only has a ln button (for natural logarithms) and a log button (for a common logarithm of base 10). You can use the ln or log button to determine $\log_2 5$ on your calculator by using the change of base formula. For the ln button, you would use $b = 2$, $x = 5$, and natural logarithms ($a = e$):

$$\log_2 5 = \frac{\ln 5}{\ln 2} \approx \frac{1.609437912}{0.693147181} \approx 2.32193$$

For the log button, you would use $b = 2$, $x = 5$, and $a = 10$:

$$\log_2 5 = \frac{\log_{10} 5}{\log_{10} 2} \approx \frac{0.698970004}{0.301029996} \approx 2.32193$$

These are applications of the change of base formula:

$$\log_b x = \frac{\log_a x}{\log_a b}$$

Example 1. Use a calculator to determine $\log_7 4$.

To use the ln button, use $b = 7$, $x = 4$, and natural logarithms ($a = e$):

$$\log_7 4 = \frac{\log_a x}{\log_a b} = \frac{\ln 4}{\ln 7} \approx 0.712$$

To use the log (base 10) button, use $b = 7$, $x = 4$, and $a = 10$:

$$\log_7 4 = \frac{\log_a x}{\log_a b} = \frac{\log_{10} 4}{\log_{10} 7} \approx 0.712$$

Example 2. Use a calculator to determine $\log_6 0.3$.

To use the ln button, use $b = 6$, $x = 0.3$, and natural logarithms ($a = e$):

$$\log_6 0.3 = \frac{\log_a x}{\log_a b} = \frac{\ln 0.3}{\ln 6} \approx -0.672$$

To use the log (base 10) button, use $b = 6$, $x = 0.3$, and $a = 10$:

$$\log_6 0.3 = \frac{\log_a x}{\log_a b} = \frac{\log_{10} 0.3}{\log_{10} 6} \approx -0.672$$

Logarithms and Exponentials Essential Skills Practice Workbook with Answers

Exercise Set 4.3

Directions: Use a calculator to determine each answer to 3 significant figures.

1) $\log_2 9 \approx$

2) $\log_{11} 3 \approx$

3) $\log_4 0.71 \approx$

4) $\log_3 \left(\frac{1}{2}\right) \approx$

5) $\log_7 42 \approx$

6) $\log_6 813 \approx$

7) $\log_5 37.5 \approx$

8) $\log_{12} 2.107 \approx$

9) $\log_{63} 2 \approx$

10) $\log_9 \left(\frac{3}{4}\right) \approx$

11) $\log_2 0.007 \approx$

12) $\log_8 1{,}000{,}000 \approx$

4.4 Logarithm Exercises Involving a Calculator

The problems of this section can be solved using the ln or log function of a calculator.

Example 1. Use a calculator to determine $\log_5 \sqrt{2}$.
To use the ln button, use $b = 5$, $x = \sqrt{2}$, and natural logarithms ($a = e$):

$$\log_5 \sqrt{2} = \frac{\log_a x}{\log_a b} = \frac{\ln \sqrt{2}}{\ln 5} \approx 0.215$$

To use the log (base 10) button, use $b = 5$, $x = \sqrt{2}$, and $a = 10$:

$$\log_5 \sqrt{2} = \frac{\log_a x}{\log_a b} = \frac{\log_{10} \sqrt{2}}{\log_{10} 5} \approx 0.215$$

Example 2. Use a calculator to determine $\log_2 \left(\frac{\pi}{3}\right)$.
To use the ln button, use $b = 2$, $x = \frac{\pi}{3}$, and natural logarithms ($a = e$):

$$\log_2 \left(\frac{\pi}{3}\right) = \frac{\log_a x}{\log_a b} = \frac{\ln \left(\frac{\pi}{3}\right)}{\ln 2} \approx 0.0665$$

To use the log (base 10) button, use $b = 2$, $x = \frac{\pi}{3}$, and $a = 10$:

$$\log_2 \left(\frac{\pi}{3}\right) = \frac{\log_a x}{\log_a b} = \frac{\log_{10} \left(\frac{\pi}{3}\right)}{\log_{10} 2} \approx 0.0665$$

Exercise Set 4.4

Directions: Use a calculator to determine each answer to 3 significant figures.

1) $\log_3 (9.8^5) \approx$

2) $\log_7 e \approx$

Logarithms and Exponentials Essential Skills Practice Workbook with Answers

3) $\log_{15}\left(\dfrac{42^7}{8^3}\right) \approx$

4) $\log_3 \sqrt{375} \approx$

5) $\sqrt{\log_6 248} \approx$

6) $\log_5\left(\dfrac{\pi^2}{3}\right) \approx$

7) $\log_8\left[6^{(5^3)}\right] \approx$

8) $\sqrt{\log_4(39^2)} \approx$

9) $\log_5\left(\dfrac{1}{e^{16}}\right) \approx$

10) $\log_9(\log_3 6) \approx$

11) $\log_6(\sin 72°) \approx$

12) $\log_2(50!) \approx$

5 Exponentials

5.1 The Exponential Function

The exponential function raises Euler's number to a variable power: e^x. Recall that Euler's number is approximately 2.71828 (Sec. 1.6). Most calculators have either an EXP button or an e^x button. Enter e^5 on your calculator and check that you get 148.4 to four significant figures.

The function e^x increases very rapidly. For example, compare the following values:
- $e^1 \approx 2.718$
- $e^{10} \approx 2.203 \times 10^4$
- $e^{100} \approx 2.688 \times 10^{43}$

In contrast, the function e^{-x} decreases vary rapidly. For example:
- $e^{-1} \approx 0.3679$
- $e^{-10} \approx 4.540 \times 10^{-5}$
- $e^{-100} \approx 3.720 \times 10^{-44}$

The term exponential growth refers to the function e^x, whereas the term exponential decay refers to the function e^{-x}. We will explore exponential growth and decay visually in Chapter 7 (and their applications in Chapter 8). In the current chapter, we will focus on performing calculations or solving equations with these functions.

All of the rules regarding exponents from Chapter 1 also apply to exponentials, such as:
$$e^x e^y = e^{x+y} \quad , \quad e^0 = 1 \quad , \quad e^{-1} = \frac{1}{e} \quad , \quad e^{-x} = \frac{1}{e^x}$$
$$(e^x)^a = e^{ax} \quad , \quad e^{1/2} = \sqrt{e} \quad , \quad e^{1/n} = \sqrt[n]{e}$$

Recall the cancellation equations from Chapter 3 (and the other rules for logarithms):
$$y = \ln(e^y) \quad , \quad x = e^{\ln x} \quad , \quad \ln(e) = 1$$

Note: One way to find e on your calculator is to enter e^1.

Logarithms and Exponentials Essential Skills Practice Workbook with Answers

Example 1. $e^3 \approx 20.1$

Example 2. $e^{-3} \approx 0.0498$

Exercise Set 5.1

Directions: Use a calculator to determine each answer to 3 significant figures.

1) $e^2 \approx$

2) $e^{-2} \approx$

3) $e^{1/4} \approx$

4) $e^{-1/4} \approx$

5) $e^{\sqrt{7}} \approx$

6) $e^e \approx$

7) $e^\pi \approx$

8) $e^{42} \approx$

9) $e^{-75} \approx$

10) $\sqrt{e} \approx$

11) $2^e \approx$

12) $2^{-e} \approx$

13) $\ln(2e) \approx$

14) $\ln\left(\dfrac{e}{2}\right) \approx$

15) $\sqrt{e^2 - 1} \approx$

16) $\dfrac{1+e}{1-e} \approx$

17) $e^{\sin 60°} \approx$

18) $e^{\sin 300°} \approx$

5.2 Solving Equations with Exponential Functions

Many equations involving exponentials and one variable can be solved as follows:
- If there is more than one term involving an exponential, see if you can combine like terms or factor out an exponential in order to reduce the number of terms that involve exponentials.
- Attempt to isolate the exponential term with the variable. If you can get the exponential term with the variable all by itself one side of the equation, like $e^x = c$, then you can take the natural logarithm of both sides of the equation. For example, $e^x = c$ becomes $\ln(e^x) = \ln c$. Apply the cancellation equation ($\ln e^x = x$) to remove the logarithm: $x = \ln c$. The examples that follow illustrate this idea.

Example 1. $3e^x + 6 = 18$
- Subtract 6 from both sides: $3e^x = 12$
- Divide by 3 on both sides: $e^x = 4$ (The exponential is now isolated.)
- Take the natural logarithm of both sides: $\ln(e^x) = \ln 4$
- Apply the cancellation equation: $x = \ln 4$
- Use a calculator to find the answer: $x \approx 1.386$

Check the answer: Plug $x \approx 1.386$ into the original equation. Since $3e^x + 6 \approx 3(e^{1.386}) + 6 \approx 3(3.999) + 6 \approx 12 + 6 = 18$, the answer checks out.

Example 2. $7e^x = 5e^x + 1$
- Subtract $5e^x$ from both sides: $2e^x = 1$
- Divide by 2 on both sides: $e^x = \frac{1}{2}$
- Take the natural logarithm of both sides: $\ln(e^x) = \ln\left(\frac{1}{2}\right)$
- Apply the cancellation equation: $x = \ln\left(\frac{1}{2}\right)$
- Use a calculator to find the answer: $x \approx -0.6931$

Check the answer: Plug $x \approx -0.6931$ into the original equation. Since $7e^x \approx 7e^{-0.6931} \approx 3.500$ and $5e^x + 1 \approx 5e^{-0.6931} + 1 \approx 2.500 + 1 = 3.500$, the answer checks out.

Exercise Set 5.2

Directions: Solve for the unknown in each equation. You may use a calculator.

1) $8e^x - 12 = 60$

2) $28 - 2e^x = 4$

3) $9e^{-x} + 18 = 54$

4) $e^{3x+2} = 4$

5) $e^{4x}e^3 = 25$

5 Exponentials

6) $e^{2x}e^x = \dfrac{1}{100}$

7) $\dfrac{12}{e^{x-2}} = 3$

8) $e^{2x} + 21 = 10e^x$

9) $e^{(e^x)} = 1000$

10) $e^{\cos x} = \dfrac{1}{2}$ (Assume that the answer lies in Quadrant II.)

5.3 Variable Exponents of Other Bases

Consider a constant base raised to a variable power, such as b^x, b^{-x}, or even b^{ax}. If the base happens to be Euler's number, we get the exponential function, e^x. However, in this section and the next we will focus on other bases, such as 2^x or 37^x. In this section, we will evaluate such functions for specific values of x. In Sec. 5.4, we will solve equations with variable exponents.

Example 1. $3^e \approx 19.8$

Example 2. $15^{-2.4} \approx 0.00150$

Exercise Set 5.3

Directions: Use a calculator to determine each answer to 3 significant figures.

1) $4^{7.2} \approx$

2) $7^{2/3} \approx$

3) $25^{-6.9} \approx$

4) $6^{-3/4} \approx$

5) $5^{\sqrt{3}} \approx$

6) $9^\pi \approx$

7) $\pi^\pi \approx$

8) $8^{\sqrt{5}} \approx$

9) $\left(\frac{4}{7}\right)^{3/5} \approx$

10) $\sqrt{e^\pi} \approx$

11) $\sqrt{7}^{\sqrt{11}} \approx$

12) $\sqrt{7^{11}} \approx$

5.4 Solving Equations with Variable Exponents

Many equations involving variable exponents can be solved as follows:
- First attempt to isolate the term with the variable exponent, like in Sec. 5.2.
- If you can isolate an expression of the form b^x, then you can take a logarithm (with matching base, b) of both sides of the equation. Apply the cancellation equation ($\log_b b^x = x$) to remove the logarithm. For example, $b^x = c$ leads to $x = \log_b c$. Use the change of base formula (Chapter 4) to compute the logarithm. The example that follows illustrates this idea.

$$\log_b x = \frac{\ln x}{\ln b} \quad \text{or} \quad \log_b x = \frac{\log_{10} x}{\log_{10} b}$$

Example 1. $2^x + 1 = 6$
- Subtract 1 from both sides: $2^x = 5$
- Take a base-2 logarithm of both sides: $\log_2(2^x) = \log_2 5$
- Apply the cancellation equation: $x = \log_2 5$
- Use the change of base formula: $x = \frac{\ln 5}{\ln 2}$ or $\frac{\log_{10} 5}{\log_{10} 2}$
- Use a calculator to find the answer: $x \approx 2.322$

Check the answer: Plug $x \approx 2.322$ into the original equation. Since $2^x + 1 \approx 2^{2.322} + 1 \approx 5 + 1 = 6$, the answer checks out.

Exercise Set 5.4

Directions: Solve for the unknown in each equation. You may use a calculator.

1) $3^x - 18 = 36$

Logarithms and Exponentials Essential Skills Practice Workbook with Answers

2) $5^{-x} + 25 = 150$

3) $8^{5x-2} = \frac{2}{3}$

4) $\frac{7^{9x}}{7^{4x}} = \frac{3}{50}$

5) $2^x 4^x 8^x = 175$

6) $6^x 36^x = 1296$

7) $\dfrac{24}{6^x+1} = \dfrac{1}{2}$

8) $3^{(2^x)} = 500$

9) $5^{2x} + 24 = 11(5^x)$

10) $\dfrac{9^x+1}{5} = \dfrac{3^x}{2}$

6 Hyperbolic Functions

6.1 Hyperbolic Sine and Cosine

The **hyperbolic functions** are expressions involving the exponential functions e^x and e^{-x} which arise in many applications of mathematics. The hyperbolic functions are named after the basic trigonometric functions because they have a similar relationship with the hyperbola as the ordinary trigonometric functions have with the circle. The hyperbolic functions add the letter "h" to the name of the corresponding trigonometric function. For example, sinh x and cosh x are the counterparts of sin x and cos x.

Hyperbolic sine (sinh x) is defined as one-half of the difference between e^x and e^{-x}:

$$\sinh x = \frac{e^x - e^{-x}}{2}$$

Hyperbolic cosine (cosh x) is defined as one-half of the sum of e^x and e^{-x}:

$$\cosh x = \frac{e^x + e^{-x}}{2}$$

Example 1. Evaluate cosh(3) without using the cosh function of a calculator.
Plug $x = 3$ into the definition of hyperbolic cosine. Use the EXP or e^x function of a calculator to determine the answer. (The problem states not to use the cosh function, but doesn't state not to use the exponential function.) Determine e^3 and e^{-3} on the calculator to calculate the answer.

$$\cosh 3 = \frac{e^3 + e^{-3}}{2} \approx 10.07$$

Exercise Set 6.1

Directions: Evaluate each of the following without using the sinh or cosh functions of a calculator. When necessary, you may use the EXP or e^x functions of a calculator. Round any inexact answers to four significant figures.

1) $\sinh 0 \approx$

2) $\cosh 0 \approx$

3) $\sinh 1 \approx$

4) $\cosh 1 \approx$

5) $\sinh 2 \approx$

6) $\cosh 2 \approx$

7) $\sinh\left(\frac{1}{2}\right) \approx$

8) $\cosh\left(\frac{1}{2}\right) \approx$

9) $\sinh(-1) \approx$

10) $\cosh(-1) \approx$

11) $\sinh(-2) \approx$

12) $\cosh(-2) \approx$

13) $\sinh(\ln 2) \approx$

14) $\cosh(\ln 2) \approx$

15) $\sinh \sqrt{2} \approx$

16) $\cosh \sqrt{2} \approx$

6.2 Hyperbolic Sine and Cosine Identities

There are several handy identities involving hyperbolic sine and hyperbolic cosine which can be proven by applying the definitions of these functions from Chapter 1.

Example 1. Prove that $\sinh x + \cosh x = e^x$.
Use the definitions of hyperbolic sine and hyperbolic cosine.
$$\sinh x + \cosh x = \frac{e^x - e^{-x}}{2} + \frac{e^x + e^{-x}}{2} = \frac{e^x - e^{-x} + e^x + e^{-x}}{2} = \frac{2e^x}{2} = e^x$$

Exercise Set 6.2

Directions: Prove each of the following identities.

1) $\cosh x - \sinh x = e^{-x}$

2) $\sinh(-x) = -\sinh x$

3) $\cosh(-x) = \cosh x$

4) $\sinh(x + y) = \sinh x \cosh y + \cosh x \sinh y$

5) $\cosh(x+y) = \cosh x \cosh y + \sinh x \sinh y$

6) $\sinh(2x) = 2 \sinh x \cosh x$

7) $\cosh(2x) = \cosh^2 x + \sinh^2 x$

8) $\cosh^2 x - \sinh^2 x = 1$

9) $(\cosh x + \sinh x)^n = \cosh(nx) + \sinh(nx)$

6.3 Other Hyperbolic Functions

The **hyperbolic secant** (sech x), **hyperbolic cosecant** (csch x), **hyperbolic tangent** (tanh x), and **hyperbolic cotangent** (coth x) functions are defined similar to the ordinary trig functions with similar names:

$$\text{sech}\, x = \frac{1}{\cosh x} = \frac{2}{e^x + e^{-x}}$$

$$\text{csch}\, x = \frac{1}{\sinh x} = \frac{2}{e^x - e^{-x}}$$

$$\tanh x = \frac{\sinh x}{\cosh x} = \frac{e^x - e^{-x}}{e^x + e^{-x}}$$

$$\coth x = \frac{\cosh x}{\sinh x} = \frac{e^x + e^{-x}}{e^x - e^{-x}}$$

Example 1. Evaluate tanh(4) without using a hyperbolic function on a calculator.
Plug $x = 4$ into the definition of hyperbolic tangent. Use the EXP or e^x function of a calculator to determine the answer. (The problem states not to use the tanh function, but doesn't state not to use the exponential function.) Determine e^4 and e^{-4} on the calculator to calculate the answer.

$$\tanh 4 = \frac{\sinh 4}{\cosh 4} = \frac{e^4 - e^{-4}}{e^4 + e^{-4}} \approx 0.9993$$

Example 2. Evaluate sech $\left(\frac{2}{3}\right)$ without using a hyperbolic function on a calculator.
Plug $x = \frac{2}{3}$ into the definition of hyperbolic secant. Determine $e^{2/3}$ and $e^{-2/3}$ on the calculator to calculate the answer.

$$\text{sech}\left(\frac{2}{3}\right) = \frac{1}{\cosh(2/3)} = \frac{2}{e^{2/3} + e^{-2/3}} \approx 0.8126$$

Note: If you were to use a calculator that has a cosh function (or a hyp function that you can access to transform cos into cosh), but which doesn't have a sech button, you would first find cosh and then find the reciprocal of that (perhaps by pressing an x^{-1} button). It would be incorrect to use a $\cosh^{-1} x$ function (which is an inverse function, which is much different from a reciprocal function, as we'll see in Sec. 6.5).

Exercise Set 6.3

Directions: Evaluate each of the following without using the hyperbolic functions of a calculator. When necessary, you may use the EXP or e^x functions of a calculator. Round any inexact answers to four significant figures.

1) $\tanh 0 \approx$

2) $\text{sech } 0 \approx$

3) $\tanh 1 \approx$

4) $\coth 1 \approx$

5) $\text{sech } 2 \approx$

6) $\text{csch } 2 \approx$

7) $\tanh 2 \approx$

8) $\coth 2 \approx$

9) $\text{sech}\left(\frac{1}{2}\right) \approx$

10) $\text{csch}\left(\frac{1}{2}\right) \approx$

11) $\tanh\left(\frac{1}{2}\right) \approx$

12) $\coth\left(\frac{1}{2}\right) \approx$

13) $\text{sech}(\ln 2) \approx$

14) $\text{csch}(\ln 2) \approx$

15) $\tanh \sqrt{2} \approx$

16) $\coth \sqrt{2} \approx$

6.4 Other Hyperbolic Identities

Hyperbolic identities involving hyperbolic secant, cosecant, tangent, and cotangent can be proven by applying the definitions from Sec. 6.3 (and also from Sec. 6.1), or by applying identities from the exercises (or example) of Sec. 6.2.

Example 1. Prove that $1 - \tanh^2 x = \text{sech}^2 x$.
Use the definitions of hyperbolic tangent and hyperbolic secant.
Method 1: Express $\tanh x$ in terms of $\sinh x$ and $\cosh x$ and apply the identity from Exercise 8 ($\cosh^2 x - \sinh^2 x = 1$) in Sec. 6.2.

$$1 - \tanh^2 x = 1 - \frac{\sinh^2 x}{\cosh^2 x} = \frac{\cosh^2 x}{\cosh^2 x} - \frac{\sinh^2 x}{\cosh^2 x}$$

$$= \frac{\cosh^2 x - \sinh^2 x}{\cosh^2 x} = \frac{1}{\cosh^2 x} = \text{sech}^2 x$$

Method 2: Express $\tanh x$ in terms of exponential functions.

$$1 - \tanh^2 x = 1 - \left(\frac{e^x - e^{-x}}{e^x + e^{-x}}\right)^2 = 1 - \frac{e^{2x} - 2e^x e^{-x} + e^{-2x}}{e^{2x} + 2e^x e^{-x} + e^{-2x}}$$

$$= 1 - \frac{e^{2x} - 2 + e^{-2x}}{e^{2x} + 2 + e^{-2x}} = \frac{e^{2x} + 2 + e^{-2x}}{e^{2x} + 2 + e^{-2x}} - \frac{e^{2x} - 2 + e^{-2x}}{e^{2x} + 2 + e^{-2x}}$$

$$= \frac{e^{2x} + 2 + e^{-2x} - (e^{2x} - 2 + e^{-2x})}{e^{2x} + 2 + e^{-2x}} = \frac{e^{2x} + 2 + e^{-2x} - e^{2x} + 2 - e^{-2x}}{e^{2x} + 2 + e^{-2x}}$$

$$= \frac{2 + 2}{e^{2x} + 2 + e^{-2x}} = \frac{4}{e^{2x} + 2 + e^{-2x}} = \left(\frac{2}{e^x + e^{-x}}\right)^2 = \text{sech}^2 x$$

Note that $e^x e^{-x} = e^{x-x} = e^0 = 1$.

Exercise Set 6.4

Directions: Prove each of the following identities.

1) $\tanh(-x) = -\tanh x$

2) $\text{sech}(-x) = \text{sech}\, x$

3) $\operatorname{csch}(-x) = -\operatorname{csch} x$

4) $\coth^2 x - 1 = \operatorname{csch}^2 x$

5) $\tanh(x+y) = \dfrac{\tanh x + \tanh y}{1 + \tanh x \tanh y}$

6) $\tanh(2x) = \dfrac{2\tanh x}{1 + \tanh^2 x}$

7) $\tanh\left(\dfrac{x}{2}\right) = \dfrac{\sinh x}{1+\cosh x}$

8) $\tanh(\ln x) = \dfrac{x^2-1}{x^2+1}$

9) $\dfrac{1+\tanh x}{1-\tanh x} = e^{2x}$

6.5 Inverse Hyperbolic Functions

Every hyperbolic function has an inverse hyperbolic function. We put a small -1 to the top right of the function's name to indicate an inverse, such as $\cosh^{-1} x$. This -1 means the inverse function. It isn't an exponent; it isn't a reciprocal; it doesn't mean to take a reciprocal.

So what does the inverse mean? Let's begin by thinking about what the hyperbolic functions (the ordinary ones, not the inverse functions yet) do. For example, $\cosh 2$ means to plug $x = 2$ into $\frac{e^x + e^{-x}}{2}$. When we do this, we get $\cosh 2 = \frac{e^2 + e^{-2}}{2} \approx 3.762$. The inverse hyperbolic cosine does the opposite in the following sense: $\cosh^{-1} 3.762$ asks the question, which number can you take the hyperbolic cosine of and obtain 3.762 as the result? The answer is $\cosh^{-1} 3.762 = 2$ because $\cosh 2 = 3.762$. The quantity $\cosh^{-1} 3.762$ asks, which value of x would make $\frac{e^x + e^{-x}}{2}$ equal 3.762.

There is another way to look at it. If you apply a hyperbolic function and then apply the corresponding inverse hyperbolic function (or vice-versa), the two effects cancel out. For example, $\cosh^{-1}(\cosh x) = x$ and $\cosh(\cosh^{-1} x) = x$. For example, if $x = 2$, we get $\cosh^{-1}(\cosh 2) \approx \cosh^{-1}(3.762) \approx 2$.

It is generally convenient to use a calculator to determine the values of the inverse hyperbolic functions. First, you need to find a calculator that has this function. Every brand of calculator works differently, and even different models of the same brand can work differently. In some models, you can access the inverse hyperbolic function through a menu. Some models may have $\sinh^{-1}$, $\cosh^{-1}$, and $\tanh^{-1}$ buttons (look closely to ensure that the button includes the letter 'h'). Some models require that you first press hyp (or 2nd and then hyp) before pressing the $\sin^{-1}$, $\cos^{-1}$, or $\tan^{-1}$ buttons (such that pressing hyp first causes the calculator to include an 'h' after $\sin^{-1}$, $\cos^{-1}$, or $\tan^{-1}$ when it is used, transforming it into $\sinh^{-1}$, $\cosh^{-1}$, or $\tanh^{-1}$). Note that it would be **<u>incorrect</u>** to use the x^{-1} button or to find a reciprocal; these are inverse functions, **<u>not</u>** reciprocal functions.

Logarithms and Exponentials Essential Skills Practice Workbook with Answers

Note: If your calculator doesn't have sech^{-1}, csch^{-1}, and coth^{-1} functions, you can determine these values by taking the $\cosh^{-1}$, $\sinh^{-1}$, or $\tanh^{-1}$ (respectively) of the reciprocal of the argument. For example, $\text{sech}^{-1} 0.4 = \cosh^{-1}\left(\frac{1}{0.4}\right)$. The reason this works is that $\text{sech}\, x = \frac{1}{\cosh x}$. You can check that $\text{sech}^{-1} 0.4 = \cosh^{-1}\left(\frac{1}{0.4}\right) \approx 1.568$ because $\cosh(1.568) = \frac{e^{1.568} + e^{-1.568}}{2} \approx 2.5 = \frac{1}{0.4}$ and $\text{sech}(1.568) = \frac{2}{e^{1.568} + e^{-1.568}} \approx 0.4$.

Note: Even functions ($\cosh^{-1}$ and sech^{-1}) have two answers ($\pm$) because $\cosh(-x) = x$ and $\text{sech}(-x) = x$. For example, $\cosh^{-1}(1.2) \approx \pm 0.6224$. Odd inverse functions have the same sign as the argument: Compare $\sinh^{-1}(-1) \approx -0.8814$ to $\sinh^{-1} 1 \approx 0.8814$.

Example 1. $\cosh^{-1} 3 \approx \pm 1.763$ **Example 2.** $\text{csch}^{-1} 0.6 = \sinh^{-1}\left(\frac{1}{0.6}\right) \approx 1.284$

Exercise Set 6.5

Directions: Use a calculator to determine each answer to 4 significant figures.

1) $\cosh^{-1} 4 \approx$

2) $\sinh^{-1} 2 \approx$

3) $\tanh^{-1} 0.5 \approx$

4) $\text{sech}^{-1} 0.5 \approx$

5) $\text{csch}^{-1} 0.75 \approx$

6) $\coth^{-1} 10 \approx$

7) $\sinh^{-1}\left(-\frac{1}{4}\right) \approx$

8) $\text{sech}^{-1}\left(\frac{4}{5}\right) \approx$

9) $\tanh^{-1}\left(\frac{2}{3}\right) \approx$

10) $\cosh^{-1}\left(\frac{5}{2}\right) \approx$

11) $\text{csch}^{-1}\left(\frac{3}{7}\right) \approx$

12) $\tanh^{-1}\left(-\frac{3}{4}\right) \approx$

13) $\coth^{-1} \sqrt{3} \approx$

14) $\cosh^{-1} \sqrt{2} \approx$

15) $\sinh^{-1} e \approx$

16) $\text{csch}^{-1} \pi \approx$

6.6 Solving Equations with Hyperbolic Functions

Many equations involving hyperbolic functions can be solved as follows:
- First attempt to isolate the term with the hyperbolic function.
- **Tip**: It may help to recall definitions and identities from Sec.'s 6.1-6.4.
- If the hyperbolic function is isolated, apply the corresponding inverse hyperbolic function to both sides of the equation, like the examples below.
    - For $\cosh x = 3$, take the inverse hyperbolic cosine of both sides to get $\cosh^{-1}(\cosh x) = \cosh^{-1} 3$, which simplifies to $x = \cosh^{-1} 3$.
    - For $\tanh x = \frac{1}{4}$, take the inverse hyperbolic tangent of both sides to get $\tanh^{-1}(\tanh x) = \tanh^{-1} 4$, which simplifies to $x = \tanh^{-1} 4$.
    - For $\cosh^{-1} x = 1.5$ (note the difference from the previous examples), take the hyperbolic cosine of both sides to get $\cosh(\cosh^{-1} x) = \cosh 1.5$, which simplifies to $x = \cosh 1.5$.

Example 1. $2.4 + \sinh x = 3.2$
- Subtract 2.4 from both sides: $\sinh x = 0.8$
- Take the inverse hyperbolic sine of both sides: $\sinh^{-1}(\sinh x) = \sinh^{-1} 0.8$
- Apply the cancellation equation: $x = \sinh^{-1} 0.8$
- Use a calculator to find the answer: $x \approx 0.7327$

Check the answer: Plug $x \approx 0.7327$ into the original equation. Since $2.4 + \sinh 0.7327 \approx 2.4 + 0.8 = 3.2$, the answer checks out.

Example 2. $\tanh^{-1} x - 0.25 = 0.35$
- Add 0.25 to both sides: $\tanh^{-1} x = 0.6$
- Take the hyperbolic tangent of both sides: $\tanh(\tanh^{-1} x) = \tanh 0.6$
- Apply the cancellation equation: $x = \tanh 0.6$
- Use a calculator to find the answer: $x \approx 0.5370$

Check the answer: Plug $x \approx 0.5370$ into the original equation. Since $\tanh^{-1} 0.5370 - 0.25 \approx 0.60 - 0.25 = 0.35$, the answer checks out.

Exercise Set 6.6

Directions: Solve for the unknown in each equation. You may use a calculator.

1) $5\cosh x - 2 = 6$

2) $\tanh^2 x = \frac{3}{4}$

3) $\sinh \sqrt{x} = 1.5$

4) $4\sinh^{-1} x = 7$

5) $8 \sinh x = 5 \cosh x$

6) $\sinh^2 x + \cosh^2 x = 5$

7) $\text{sech}^2 x + \text{csch}^2 x + \tanh^2 x = 1.6$

8) $\sinh^{-1}(\cosh x) = 2$

6.7 Relating Hyperbolic Functions to Logarithms

Since the hyperbolic functions are defined in terms of e^x (such as $\cosh x = \frac{e^x + e^{-x}}{2}$) and since the natural logarithm is the inverse of the exponential function in the sense that $e^{\ln x} = x$ and $\ln(e^x) = x$, it may come as no surprise that each of the inverse hyperbolic functions can be expressed in terms of natural logarithms.

$$\sinh^{-1} x = \ln\left(x + \sqrt{x^2 + 1}\right) \text{ for all } x$$

$$\cosh^{-1} x = \ln\left(x + \sqrt{x^2 - 1}\right) \text{ for } x \geq 1$$

$$\tanh^{-1} x = \frac{1}{2}\ln\left(\frac{1+x}{1-x}\right) \text{ for } -1 < x < 1 \text{ (or } |x| < 1)$$

$$\coth^{-1} x = \frac{1}{2}\ln\left(\frac{x+1}{x-1}\right) \text{ for } 1 < |x|$$

$$\operatorname{sech}^{-1} x = \ln\left(\frac{1}{x} + \sqrt{\frac{1}{x^2} - 1}\right) \text{ for } 0 < x \leq 1$$

$$\operatorname{csch}^{-1} x = \ln\left(\frac{1}{x} + \sqrt{\frac{1}{x^2} + 1}\right) \text{ for } x \neq 0$$

Why are the values of x for these inverse hyperbolic functions restricted? It has to do with the possible values of the corresponding hyperbolic functions. The **domain** of the inverse function (which refers to the possible values of x for its argument) corresponds to the **range** of the function (which refers to the possible values of the function). For example, the function $\cosh x = \frac{e^x + e^{-x}}{2}$ has values that lie in the range $\cosh x \geq 1$ (since the numerator, $e^x + e^{-x}$, will never be less than the denominator, 2). Therefore, its inverse function, $\cosh^{-1} x$, has a domain of $x \geq 1$. If we express the function as $y = \cosh x$ and invert this, we get the inverse function in the form $x - \cosh^{-1} y$. Observe that the variable y corresponds to the range of the function $y = \cosh x$ and the domain of the inverse function $x = \cosh^{-1} y$.

For the inverses of the even functions, $\cosh^{-1} x$ and $\operatorname{sech}^{-1} x$ (note that $\cosh x$ and $\operatorname{sech} x$ are even functions because changing the sign of the argument doesn't change the value of the function; see Problem 3 in Sec. 6.2 and Problem 2 in Sec. 6.4), the above formulas provide only the positive solutions.

Similarly, the range of $\text{sech}\, x = \frac{1}{\cosh x} = \frac{2}{e^x + e^{-x}}$ and the domain of $\text{sech}^{-1} x$ are $0 < x \leq 1$ (the denominator, $e^x + e^{-x}$, can't be smaller than the numerator, 2), the range of $\tanh x = \frac{\sinh x}{\cosh x} = \frac{e^x - e^{-x}}{e^x + e^{-x}}$ and the domain of $\tanh^{-1} x$ are $-1 < x < 1$ (the numerator, $e^x - e^{-x}$, is always smaller than the denominator, $e^x + e^{-x}$), the range of $\coth x = \frac{\cosh x}{\sinh x} = \frac{e^x + e^{-x}}{e^x - e^{-x}}$ and the domain of $\coth^{-1} x$ are $x < -1$ and $1 < x$ which can be concisely combined into $1 < |x|$ (the numerator, $e^x + e^{-x}$, is always larger than the denominator, $e^x - e^{-x}$), and the range of $\text{csch}\, x = \frac{1}{\sinh x} = \frac{2}{e^x - e^{-x}}$ and the domain of $\text{csch}^{-1} x$ exclude the point where x would equal zero because the denominator, $e^x - e^{-x}$, would result in division by zero at that point (which is undefined).

For $\text{sech}^{-1} x$, note that $\frac{1}{x} + \sqrt{\frac{1}{x^2} - 1} = \frac{1}{x} + \sqrt{\frac{1}{x^2} - \frac{x^2}{x^2}} = \frac{1}{x} + \sqrt{\frac{1-x^2}{x^2}} = \frac{1}{x} + \frac{\sqrt{1-x^2}}{x}$, such that $\text{sech}^{-1} x$ can alternatively be expressed as $\text{sech}^{-1} x = \ln\left(\frac{1}{x} + \frac{\sqrt{1-x^2}}{x}\right)$. If you try to do this with $\text{csch}^{-1} x$, you will run into a problem. The problem is that for $\text{csch}^{-1} x$ (unlike $\text{sech}^{-1} x$), the value of x may be negative in the expression $\frac{1}{x} + \sqrt{\frac{1}{x^2} + 1}$. That's because the domain of $\text{csch}^{-1} x$ is $x \neq 0$ (which includes negative values), whereas the domain of $\text{sech}^{-1} x$ is $0 < x \leq 1$ (which is nonnegative). When we write $\frac{1}{x} + \sqrt{\frac{1}{x^2} + 1}$, the first term may be positive or negative whereas the second term is strictly positive. If we were to rewrite this as $\frac{1}{x} + \frac{\sqrt{1+x^2}}{x}$, both terms would have the same sign. This causes a problem for $\text{csch}^{-1} x$, which may have negative x, but the similar idea isn't a problem for $\text{sech}^{-1} x$, which can't have negative x. If you look up the equations that relate the inverse hyperbolic functions to natural logarithms online or in textbooks, you may find the alternate form of $\text{sech}^{-1} x$ mentioned in this paragraph instead of the one listed on the previous page.

Logarithms and Exponentials Essential Skills Practice Workbook with Answers

Example 1. Derive the equation $\cosh^{-1} x = \ln(x + \sqrt{x^2 - 1})$.

- Let $y = \cosh^{-1} x$.
- Take the hyperbolic cosine of both sides: $\cosh y = x$.
- Since $\cosh y = \frac{e^y + e^{-y}}{2}$, we may rewrite the previous equation as $\frac{e^y + e^{-y}}{2} = x$.
- Multiply by 2 on both sides: $e^y + e^{-y} = 2x$.
- Subtract $2x$ from both sides: $e^y - 2x + e^{-y} = 0$.
- Multiply by e^y on both sides: $e^{2y} - 2xe^y + 1 = 0$. Note that $e^y e^y = e^{2y}$ and that $e^y e^{-y} = e^0 = 1$ because $x^m x^n = x^{m+n}$ (Chapter 1).
- $e^{2y} - 2xe^y + 1 = 0$ a quadratic equation in e^y. It has a squared term (e^{2y}), a linear term ($2xe^y$ includes e^y), and a constant term (1). Apply the formula for the quadratic equation with $a = 1$, $b = -2x$, and $c = 1$ to solve for e^y:
$$e^y = \frac{-b \pm \sqrt{b^2 - 4ac}}{2a} = \frac{-(-2x) \pm \sqrt{(-2x)^2 - 4(1)(1)}}{2(1)} = \frac{2x \pm \sqrt{4x^2 - 4}}{2}$$
- Factor the 4 out of the radical: $\sqrt{4x^2 - 4} = \sqrt{4(x^2 - 1)} = \sqrt{4}\sqrt{x^2 - 1} = 2\sqrt{x^2 - 1}$.
$$e^y = \frac{2x \pm 2\sqrt{x^2 - 1}}{2} = \frac{2x}{2} \pm \frac{2\sqrt{x^2 - 1}}{2} = x \pm \sqrt{x^2 - 1}$$
- Only the plus sign of the $\pm$ meets the domain requirement of $x \geq 1$ for $\cosh^{-1} x$, which we discussed earlier in this chapter: $e^y = x + \sqrt{x^2 - 1}$.
- Take the natural logarithm of both sides: $\ln(e^y) = \ln(x + \sqrt{x^2 - 1})$.
- Apply the cancellation equation (Chapter 3): $y = \ln(x + \sqrt{x^2 - 1})$.
- Recall that $y = \cosh^{-1} x$ from the first bullet point: $\cosh^{-1} x = \ln(x + \sqrt{x^2 - 1})$.

Exercise Set 6.7

Directions: Derive each equation.

1) $\sinh^{-1} x = \ln(x + \sqrt{x^2 + 1})$

2) $\tanh^{-1} x = \frac{1}{2} \ln\left(\frac{1+x}{1-x}\right)$

3) $\operatorname{sech}^{-1} x = \ln\left(\frac{1}{x} + \sqrt{\frac{1}{x^2} - 1}\right)$

7 Graphs

7.1 Graphs of Powers

Given an equation where y is solved in terms of x, such as $y = 2x^2 - 3$, one way to sketch a graph for the equation is to make a table of x and y for representative values of x. For example, the table below was made by plugging each value of x into $y = 2x^2 - 3$. For example, when $x = 2$, we get $y = 2x^2 - 3 = 2(2)^2 - 3 = 2(4) - 3 = 8 - 3 = 5$.

x	-2	-1	0	1	2
y	5	-1	-3	-1	5

Once you have a table of x and y values, you can plot these points on a graph and then sketch a smooth curve through the points, as illustrated in the examples that follow.

It may also be useful to investigate the domain and range of a function, and to explore the function's behavior as x approaches particular values, such as 0 or $\pm\infty$.

- The **domain** of y refers to the set of values of x for which y is real. For example, the domain of $y = \sqrt{x}$ extends from $x = 0$ to $x = \infty$; it doesn't include negative values since the square root of a negative number isn't real.
- The **range** of y refers to the real values that y can have for the values of x in its domain. For example, the range of $y = \sin x$ extends from $y = -1$ to $y = 1$ since the sine function oscillates between -1 and 1.
- To explore limiting values, consider what happens to y (based on the equation given in a problem) as x approaches a particular value. For example, $y = e^{-x}$ approaches zero as x becomes very large. When $x = 50$, we get $y = e^{-50} \approx 1.93 \times 10^{-22}$, when $x = 100$, we get $y = e^{-100} \approx 3.72 \times 10^{-44}$, and the larger x, the closer y gets to zero. As another example, $y = e^{-x}$ approaches one as x approaches zero. When $x = 0$, we get $y = e^{-0} = e^0 = 1$.

Example 1. Sketch a graph of $y = \frac{x^2}{10}$ over the interval $-10 \leq x \leq 10$.

Make a table of x and y values as x varies from -10 to 10. If we vary x in increments of 5, we'll get a handful of data points. Plug each value of x into $y = \frac{x^2}{10}$. For example, for $x = 5$ we get $y = \frac{x^2}{10} = \frac{5^2}{10} = \frac{25}{10} = 2.5$.

x	-10	-5	0	5	10
y	10	2.5	0	2.5	10

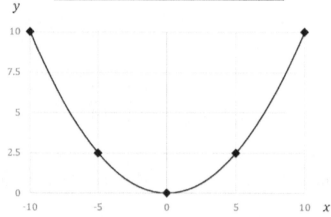

Example 2. Sketch a graph of $y = \sqrt{x}$ over the interval $0 \leq x \leq 10$.

Make a table of x and y values as x varies from 0 to 10. If we vary x in increments of 2, we'll get a handful of data points. Plug each value of x into $y = \sqrt{x}$. For example, for $x = 4$ we get $y = \sqrt{x} = \sqrt{4} = 2$. (Note that y isn't real for negative x.)

x	0	2	4	6	8	10
y	0	1.4	2	2.4	2.8	3.2

Exercise Set 7.1

Directions: Sketch a graph for each equation over the indicated interval. You may use a calculator to make the table of x and y values. (For even roots like $x^{1/2}$ or $x^{1/4}$, only graph the positive solutions.) Comment on the domain and range of each function outside of the specified interval.

1) $y = \dfrac{x^3}{8}$ for $-4 \leq x \leq 4$

2) $y = \dfrac{x^4}{16}$ for $-4 \leq x \leq 4$

3) $y = x^{3/2}$ for $0 \leq x \leq 4$

4) $y = x^{1/3}$ for $-64 \leq x \leq 64$

7.2 Graphs of Logarithms

When graphing a logarithm, note that the domain only includes values of x that aren't negative. You can't find the logarithm of a negative argument. To see this, consider the logarithm $y = \log_2 x$, which is equivalent to $2^y = x$. For example, for $x = 8$, we get $y = \log_2 8 = 3$, which agrees with $2^3 = 8$. Observe that x can't be negative because $2^y = x$ and no real value of y can make 2^y negative.

Logarithms adhere to the following limiting behavior:
- As the argument approaches 0 (from the right), the logarithm becomes more and more negative (approaching $-\infty$). Graphs of logarithms therefore feature a vertical asymptote where the argument approaches zero.
- As the argument gets very large (approaching ∞), the logarithm also gets very large (also approaching ∞), but at a very slow rate. For example, for $y = \ln x$, when $x = 1{,}000{,}000{,}000{,}000 = 10^{12}$, we get $y = \ln(10^{12}) = 27.63$, and for $x = 10^{99}$, we get $y = \ln(10^{99}) = 228$. Logarithms grow very slowly.
- As the argument approaches 1, the logarithm approaches zero. For example, $\ln 1 = 0$ because $e^0 = 1$. When the argument is smaller than 1, the logarithm is negative, and when the argument is greater than 1, the logarithm is positive.

Example 1. Sketch a graph of $y = \ln x$ over the interval $0 < x \leq 4$.

Make a table of x and y values as x varies from 0 to 4, but don't plug in $x = 0$ because $\ln x$ approaches $-\infty$ as x approaches zero. We'll explore a handful of values of x between 0 and 1 and another handful of values of x between 1 and 4. Plug each value of x into $y = \ln x$. For example, for $x = 2$, a calculator gives $y = \ln x = \ln 2 \approx 0.693$. (In the actual graph that we made on the following page, we kept more digits than are shown in the table below. This is true for all of the graphs in this book.)

x	0.2	0.4	0.6	0.8	1	2	3	4
y	-1.6	-0.9	-0.5	-0.2	0	0.7	1.1	1.4

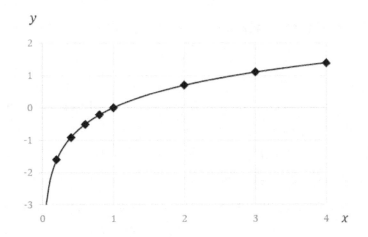

Exercise Set 7.2

Directions: Sketch a graph for each equation over the indicated interval. You may use a calculator to make the table of x and y values.

1) $y = \log_{10} x$ for $0 < x \leq 10$

2) $y = \log_2 x$ for $0 < x \leq 8$

7.3 Graphs of Exponentials and Variable Powers

When graphing an exponential function, note the following:
- The range of the exponential functions $y = e^x$ and $y = e^{-x}$ is from $y = 0$ to ∞. No value of x in either case can make y negative. (You can get negative values of y if there is an overall minus sign in front of Euler's number, like $y = -e^x$, but in that case y will never be positive.)
- The exponential function $y = e^x$ grows rapidly as x gets larger and larger, while the function $y = e^{-x}$ approaches zero as x gets larger.
- When the exponent is zero, the exponential equals one: $e^0 = 1$.

Example 1. Sketch a graph of $y = e^x$ over the interval $-5 \leq x \leq 5$.

Make a table of x and y values as x varies from -5 to 5. Plug each value of x into $y = e^x$. Use a calculator. For example, for $x = 2.5$ we get $y = e^x = e^{2.5} \approx 12.18$.

x	-5	-2.5	0	2.5	5
y	0.007	0.08	1	12	148

As x varies from -5 to 5, y varies from 0.007 to 148, which is a relatively large range (equal to roughly five powers of ten). In Sec. 7.5, we'll discuss an alternative method for graphing exponential functions that naturally accommodates such large ranges of data.

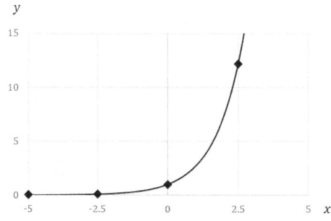

Note: The last point $(5, 148)$ would go way off the chart (so it is not shown).

Example 2. Sketch a graph of $y = e^{-x}$ over the interval $0 \leq x \leq 5$.
Make a table of x and y values as x varies from 0 to 5. Plug each value of x into $y = e^{-x}$.
Use a calculator. For example, for $x = 2$ we get $y = e^{-x} = e^{-2} \approx 0.135$.

x	0	1	2	3	4	5
y	1	0.37	0.14	0.05	0.02	0.007

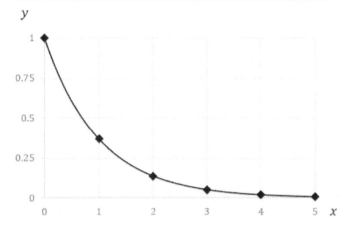

Exercise Set 7.3

Directions: Sketch a graph for each equation over the indicated interval. You may use a calculator to make the table of x and y values.

1) $y = 1 - e^{-x}$ for $0 \leq x \leq 5$

2) $y = 2^x$ for $-5 \leq x \leq 5$

3) $y = e^{(x^2)}$ for $-2 \leq x \leq 2$

4) $y = e^{-(x^2)}$ for $-4 \leq x \leq 4$

5) $y = \frac{1}{1+e^x}$ for $-5 \leq x \leq 5$

6) $y = \frac{1}{1-e^{-x}}$ for $-5 \leq x \leq 5$

7.4 Graphs of Hyperbolic Functions

As discussed in Chapter 6, each hyperbolic function is related to exponential functions. For example, $\cosh x = \frac{e^x + e^{-x}}{2}$. It may help to review Chapter 6.

Example 1. Sketch a graph of $y = \cosh x$ over the interval $-2 \leq x \leq 2$.
Recall that $\cosh x = \frac{e^x + e^{-x}}{2}$. Make a table of x and y values.

x	-2	-1	0	1	2
y	3.8	1.5	1	1.5	3.8

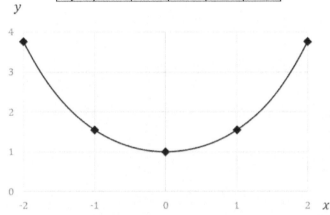

Example 2. Sketch a graph of $y = \sinh x$ over the interval $-2 \leq x \leq 2$.
Recall that $\sinh x = \frac{e^x - e^{-x}}{2}$. Make a table of x and y values.

x	-2	-1	0	1	2
y	-3.6	-1.2	0	1.2	3.6

Exercise Set 7.4

Directions: Sketch a graph for each equation over the indicated interval. You may use a calculator to make the table of x and y values.

1) $y = \text{sech}\, x$ for $-5 \leq x \leq 5$

2) $y = \text{csch}\, x$ for $-5 \leq x \leq 5$

3) $y = \tanh x$ for $-2 \leq x \leq 2$

4) $y = \coth x$ for $-2 \leq x \leq 2$

5) $y = \sinh^{-1} x$ for $-4 \leq x \leq 4$

6) $y = \cosh^{-1} x$ for $0 < x \leq 5$
(Only graph the positive root for the inverse hyperbolic cosine.)

7) $y = \tanh^{-1} x$ for $-1 < x < 1$

8) $y = \text{sech}^{-1} x$ for $0 < x \leq 1$
(Only graph the positive root for the inverse hyperbolic secant.)
Tip: Review Sec. 6.7.

7.5 Semi-Log Plots

When graphing the exponential function $y = e^x$ or other functions with a variable in the exponent like $y = 8^x$, a modest change in x can produce a dramatic change in y. For example, as x changes from 10 to 100, $y = e^x$ changes from 22,026 to 2.7×10^{43}. This makes it difficult for an ordinary graph to show all of the features of such a graph.

The solution to this problem is to use a **semi-log plot**. A semi-log plot has a vertical axis that is logarithmically scaled. Instead of scaling the vertical axis linearly in even increments like 1, 2, 3, 4, 5, etc. or like 10, 20, 30, 40, 50, etc., a semi-log plot scales the vertical axis so that powers of ten (1, 10, 100, 1000, etc.) are evenly spaced. You can see this by comparing the two graphs below, which both correspond to $y = e^x$. The top graph is an ordinary graph with linearly scaled axes, whereas the bottom graph is a semi-log plot with a logarithmically scaled vertical axis.

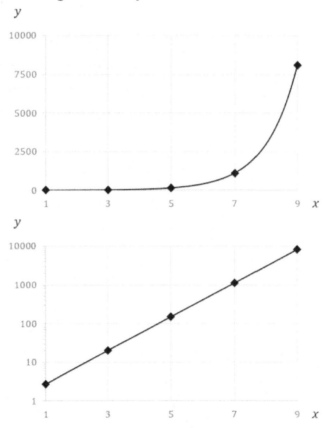

Logarithms and Exponentials Essential Skills Practice Workbook with Answers

Look closely at the tick marks on the vertical axis. Note that the nonlinear spacing of 1, 2, 3, etc., of 10, 20, 30, etc, of 100, 200, 300, etc., and so on the semi-log plot.

Observe that a semi-log plot **linearizes** the curve $y = e^x$. When we graph this function with a semi-log plot, the result is a straight line. The effect of logarithmically scaling the vertical axis is similar to applying a logarithm to y, and since $y = e^x$ for this curve, taking the logarithm of y effectively takes a logarithm of e^x, for which the cancellation equation tells us that $\ln(e^x) = x$. That is, the logarithm cancels out the exponential, transforming the "curve" into a straight line. (Technically, we scaled the vertical axis with a base-10 logarithm, but the change of base formula, $\log_{10} x = \frac{\ln x}{\ln 10}$, lets us relate a base-10 logarithm to a natural logarithm, with the only difference being a constant factor of $\frac{1}{\ln 10}$. So it doesn't matter whether we scale the vertical axis with a base-10 logarithm or natural logarithm; either way, the exponential appears as a straight line.)

Example 1. Sketch a graph of $y = e^{(x^2)}$ over the interval $0 \leq x \leq 2$.
Make a table of x and y values as x varies from 0 to 2. Plug each value of x into $y = e^{(x^2)}$. Use a calculator. For example, for $x = 2$ we get $y = e^{(x^2)} = e^{(2^2)} = e^4 \approx 55$.

x	0	0.5	1	1.5	2
y	1	1.3	2.7	9.5	55

Note that the semi-log plot of $y = e^{(x^2)}$ has the shape of a parabola (x^2). The logarithmic scaling has the effect of the cancellation equation $\ln[e^{(x^2)}] = x^2$.

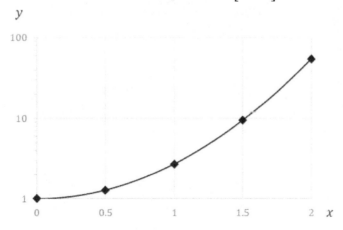

Exercise Set 7.5

Directions: Sketch a **semi-log** graph for each equation over the indicated interval. You may use a calculator to make the table of x and y values.

1) $y = e^{-x}$ for $0 \leq x \leq 8$

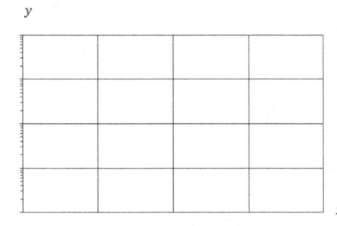

2) $y = 2^x$ for $0 \leq x \leq 10$

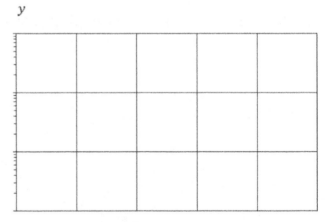

3) $y = \frac{1}{1-e^{-x}}$ for $0 < x \leq 1$

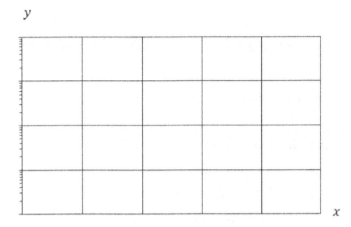

4) $y = x^2$ for $1 \leq x \leq 100$

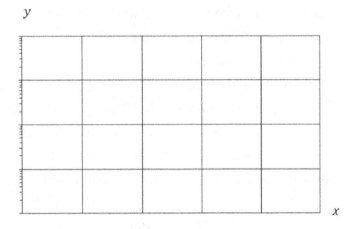

8 Applications

8.1 A Simple Model for Population Growth

A simple model for natural population growth is:
$$N = N_0 e^{kt}$$
assuming that k is constant.
- t represents time.
- N_0 is the initial population when $t = 0$.
- N is the population at time t.
- k is the relative growth rate. Its units must be the reciprocal of t's units. For example, if the time is measured in years, then k has units of $\frac{1}{\text{years}}$.

Example 1. A certain species has an initial population of 1800 organisms on an island. Every 10 years, the population increases by an average of 25%. What is the relative growth rate?

Identify the information given in the problem:
- $N_0 = 1800$ is the initial population.
- $t = 10$ years is the time.
- $N = 1800 \times 1.25\% = 1800 \times 1.25 = 2250$ is the population after 10 years. The population increases by 25% every 10 years. If you increase 1800 by 25%, you get 2250. (Note: The relative growth rate isn't 0.25. We'll figure out what the relative growth rate is from the equation.)

Plug the known values into the equation. Solve for the unknown.
$$N = N_0 e^{kt}$$
$$2250 = 1800 e^{k10} \text{ (divide both sides by 1800)}$$
$$1.25 = e^{10k} \text{ (take the natural logarithm of both sides)}$$
$$\ln(1.25) = 10k \text{ (divide both sides by 10)}$$
$$k = \frac{\ln(1.25)}{10} \approx 0.0223$$

Logarithms and Exponentials Essential Skills Practice Workbook with Answers

Exercise Set 8.1

Directions: Apply the population model of this section to answer each question.

1) A culture of bacteria has an initial population of 720 cells. Each cell splits into two cells in an average of 15 minutes.

(A) What is the relative growth rate in units of $\frac{1}{\text{hr.}}$?

(B) How many cells will the culture have after 4 hours?

(C) When will the culture have a population of one million cells?

8 Applications

8.2 Half-Life

Unstable nuclei can spontaneously decay into lighter nuclei. This is called radioactive nuclear decay. The **half-life** provides a measure of the rate at which an unstable radioactive nucleus decays. Specifically, the half-life is the time it takes for one-half of the substance to decay. For example, bismuth-210 has a half-life of 5 days. If we begin with a sample of 96 grams of bismuth-210, after 5 days there will be 48 grams of bismuth-210, after 10 days there will be 24 grams, after 15 days there will be 12 grams, after 20 days there will be 6 grams, etc. Every 5 days, one-half of the sample decays, such that the mass of bismuth-210 is cut in half every 5 days.

We can model radioactive nuclear decays with the following formula:
$$m = m_0 e^{-kt}$$

- t represents time.
- m_0 is the initial mass of the radioactive substance when $t = 0$.
- m is the mass at time t.
- k is the decay constant. Its units must be the reciprocal of t's units.

Since the half-life is the time that it takes for one-half of the sample to decay, the half-life corresponds to the time it takes for m to equal $m = \frac{m_0}{2}$. If we substitute this into the formula above, and call the time $t_{½}$ (the symbol for half-life), we get:
$$\frac{m_0}{2} = m_0 e^{-kt_{½}}$$
Divide both sides of the equation by the initial mass.
$$\frac{1}{2} = e^{-kt_{½}}$$
Take the natural logarithm of both sides.
$$\ln\left(\frac{1}{2}\right) = -kt_{½}$$
Recall from Chapter 3 that $\ln\left(\frac{1}{x}\right) = -\ln x$.
$$-\ln 2 = -kt_{½}$$

Multiply both sides by negative one.
$$\ln 2 = kt_{1/2}$$
Divide both sides by the decay constant.
$$t_{1/2} = \frac{\ln 2}{k} \approx \frac{0.693}{k}$$
The half-life and decay constant are related via the above equation. We can invert this to get:
$$k = \frac{\ln 2}{t_{1/2}} \approx \frac{0.693}{t_{1/2}}$$

Notation: Some books work with $m_0 e^{kt}$ and define k to be a negative number, whereas we are working with $m_0 e^{-kt}$ and defining k to be a positive number. The two methods are mathematically equivalent. In each case, the exponent is effectively negative.

Example 1. The half-life of carbon-14 is 5730 years. A sample initially contains 72 g of carbon-14. How much carbon-14 will remain after 20,000 years?

Identify the information given in the problem:
- $m_0 = 72$ g is the initial mass.
- $t = 20,000$ years is the time.
- $t_{1/2} = 5730$ years is the half-life.

First solve for the decay constant.
$$k = \frac{\ln 2}{t_{1/2}} = \frac{\ln 2}{5730} \approx 0.000121 \frac{1}{\text{yr.}}$$
Plug the known values into the radioactive decay formula. Solve for the unknown.
$$m = m_0 e^{-kt} = 72 e^{-0.000121(20,000)} = 72 e^{-2.42} \approx 6.4 \text{ g}$$

Exercise Set 8.2

Directions: Apply the half-life formulas of this section to answer each question.

1) A sample of thorium-234 has an initial mass of 500 g. Only 10% of the thorium-234 remains after 80 days.

(A) What is the decay constant in units of $\frac{1}{day}$?

(B) What is the half-life of thorium-234?

(C) When will only 2.5 g of thorium-234 remain?

8.3 RC Circuits

When a charged capacitor is connected to a resistor (with no battery in the circuit), the charge (Q) stored on the capacitor and the current (I) in the circuit each decay exponentially:

$$Q = Q_m e^{-t/\tau}$$
$$I = I_m e^{-t/\tau}$$

- t represents time.
- Q_m and I_m represent the maximum charge and current, respectively.
- Q and I represent the charge and current, respectively, at time t.
- τ is the **time constant**. It needs to be expressed in the same units as t. The time constant is related to the resistance (R) and capacitance (C) through $\tau = RC$. When R is expressed in Ohms (Ω) and C is expressed in Farads (F), the time constant comes out in seconds. Note: τ is the Greek letter tau.

When an initially uncharged capacitor and resistor are connected in series with a battery, the charge stored on the capacitor grows while the current in the circuit decays:

$$Q = Q_m(1 - e^{-t/\tau})$$
$$I = I_m e^{-t/\tau}$$

In this case, the potential difference (ΔV) across the battery is related to the maximum charge and current by:

$$Q_m = C\Delta V_B$$
$$\Delta V_B = I_m R$$

The potential difference comes out in Volts (V) when the charge is in Coulombs (C) and the capacitance is in Farads (F) or when the current is in Ampères (A) and the resistance is in Ohms (Ω). These are all SI units.

To relate the time constant to the half-life, set the current (or charge) equal to one-half of its maximum value and replace the time with the half-life ($t_{1/2}$).

$$\frac{I_m}{2} = I_m e^{-t_{1/2}/\tau}$$

8 Applications

Divide both sides by the maximum current.
$$\frac{1}{2} = e^{-t_{1/2}/\tau}$$
Take the natural logarithm of both sides.
$$\ln\left(\frac{1}{2}\right) = -\frac{t_{1/2}}{\tau}$$
Recall from Chapter 3 that $\ln\left(\frac{1}{x}\right) = -\ln x$.
$$-\ln 2 = -\frac{t_{1/2}}{\tau}$$
Multiply both sides by negative one.
$$\ln 2 = \frac{t_{1/2}}{\tau}$$
Multiply both sides by the time constant.
$$t_{1/2} = \tau \ln 2 \approx 0.693 t_{1/2}$$
The half-life and time constant are related via the above equation. We can invert this to get:
$$\tau = \frac{t_{1/2}}{\ln 2} \approx \frac{t_{1/2}}{0.693}$$

Example 1. A 2×10^{-4} F capacitor with an initial charge of 3.6×10^{-3} C discharges while connected to a 6 Ω resistor. What will the charge on the capacitor be after 0.005 s?

Identify the information given in the problem:
- $Q_m = 3.6 \times 10^{-3}$ C is the initial charge.
- $C = 2 \times 10^{-4}$ F is the capacitance.
- $R = 6$ Ω is the resistance.
- $t = 0.005$ s is the time.

First solve for the time constant.
$$\tau = RC = (6)(2 \times 10^{-4}) = 1.2 \times 10^{-3} \text{ s} = 0.0012 \text{ s}$$
Plug the known values into the formula for the charge of a discharging capacitor.
$$Q = Q_m e^{-t/\tau} = (3.6 \times 10^{-3})e^{-0.005/0.0012} = (3.6 \times 10^{-3})e^{-4.17} \approx 5.6 \times 10^{-5} \text{ C}$$

Logarithms and Exponentials Essential Skills Practice Workbook with Answers

Exercise Set 8.3

Directions: Apply the RC circuit equations of this section to answer each question.

1) A 6×10^{-4} F capacitor with an initial charge of 3×10^{-3} C discharges while connected to a 25 Ω resistor.

(A) What is the time constant?

(B) What is the half-life?

(C) When will the capacitor have a charge of 1×10^{-4} C?

(D) After 0.03 s, what will be the charge on the capacitor?

8 Applications

2) A 4×10^{-4} F capacitor with no initial charge is connected in series with a 10 V battery and a 50 Ω resistor.

(A) What are the maximum current and maximum charge? When is each maximum?

(B) What is the time constant?

(C) When will the charge on the capacitor be one-half of its maximum value?

(D) After 0.05 s, what will be the charge on the capacitor and the current in the circuit?

8.4 The Decibel Scale

The intensity (I) of a sound wave has SI units of Watts per square meter (W/m²). In these units, the human ear can hear a very wide range of sounds: from 10^{-12} W/m² to 1 W/m², which is a range of 12 orders of magnitude (powers of 10). To deal with this wide range, scientists work with the decibel (dB) system, which uses a logarithmic scale to measure loudness (L). In terms of decibels, the human ear can hear sounds from 0 dB to 120 dB (where each increment of 10 dB corresponds to a power of 10 of intensity).

$$L = 10 \log_{10}\left(\frac{I}{I_0}\right)$$

- $I_0 = 1 \times 10^{-12}$ W/m² is the reference intensity, corresponding to the threshold of human hearing.
- I represents the intensity of a particular sound wave in W/m².
- L is the loudness of the sound wave in decibels (dB) corresponding to I.

To solve for the intensity, first divide by 10 on both sides:

$$\frac{L}{10} = \log_{10}\left(\frac{I}{I_0}\right)$$

Exponentiate both sides to base 10. Recall the cancellation equation $b^{\log_b x} = x$ from Chapter 3. In this case, $b = 10$.

$$10^{L/10} = \frac{I}{I_0}$$

Multiply both sides of the equation by the reference intensity to get $I_0 10^{L/10} = I$, which is equivalent to:

$$I = I_0 10^{L/10}$$

Example 1. A vacuum cleaner makes a sound wave with an intensity of 1×10^{-5} W/m². How loud is this sound in decibels?

$$L = 10 \log_{10}\left(\frac{I}{I_0}\right) = 10 \log_{10}\left(\frac{1 \times 10^{-5}}{1 \times 10^{-12}}\right) = 10 \log_{10}\left[10^{-5-(-12)}\right]$$
$$= 10 \log_{10}(10^{-5+12}) = 10 \log_{10}(10^7) = 10(7) = 70 \text{ dB}$$

8 Applications

Exercise Set 8.4

Directions: Apply the sound wave formulas of this section to answer each question.

1) A particular motor makes a sound wave with an intensity of 1×10^{-3} W/m². How loud is this sound in decibels?

2) When a boy talks casually, the sound of his voice has a sound level of 50 dB. What is the intensity of this sound wave?

3) How many times louder is an 80 dB sound compared to a 30 dB sound? (That is, if you take the ratio of the two intensities, what will it be?)

8.5 Continuously Compounded Interest

Recall the formula for compound interest from Sec. 1.6: $A = P\left(1 + \frac{r}{n}\right)^{nt}$. We may rewrite this formula as $A = P\left[\left(1 + \frac{r}{n}\right)^{\frac{n}{r}}\right]^{\frac{r}{n}rt}$ according to the rule $(x^p)^q = x^{pq}$ from Sec. 1.5 since $\frac{n}{r}rt = nt$. If we define $m = \frac{n}{r}$, the formula becomes $A = P\left[\left(1 + \frac{1}{m}\right)^m\right]^{rt}$, since $\frac{1}{m} = \frac{r}{n}$. If the interest is compounded continuously, the compounding frequency (n) approaches infinity, and $m = \frac{n}{r}$ also approaches infinity. For continuously compounded interest, the formula becomes $A = P\left[\lim_{m\to\infty}\left(1 + \frac{1}{m}\right)^m\right]^{rt}$. In Sec. 1.6, we learned that $\lim_{m\to\infty}\left(1 + \frac{1}{m}\right)^m = e$, such that the formula becomes:

$$A = Pe^{rt}$$

- P represents the initial balance (it is called the **principal**).
- A represents the final balance (accounting for the principal plus the interest).
- r represents the **interest rate** (in decimal form; for example, 0.3 means 30%).
- t represents the time (typically, in years).

Example 1. If $800 is invested in an account that earns 5% interest, compounded continuously, what will the balance be after 4 years?

First divide the interest rate by 100% to convert it to a decimal: $r = \frac{5\%}{100\%} = 0.05$. The principal is $P = \$800$ and the time is $t = 4$ years.

$$A = Pe^{rt} = \$800e^{0.05(4)} = \$800e^{0.2} = \$977.12$$

Exercise Set 8.5

Directions: Apply the interest formula of this section to answer each question.

1) A person takes out a loan of $4500 at 20% interest, and will pay the full amount due in 3 years. What is the total amount due in 3 years? (Note that most loans require monthly payments to be made, whereas in this problem no payments will be made for 3 years.)

2) If an unknown amount of money is invested in an account that earns 2.5% interest, compounded continuously, when will the balance triple?

9 Calculus

9.1 Limits

The notation $\lim_{x \to c} f(x)$ means, "What value does the function $f(x)$ approach as the variable x approaches the value of c?" The notation $\lim_{x \to c^+} f(x)$ and $\lim_{x \to c^-} f(x)$ are similar, except that they specify that the direction of approach is from the right or left; these are called one-sided limits.

To find the limit of the ratio of two functions, $\lim_{x \to c} \frac{f(x)}{g(x)}$, if both functions approach zero or if both functions grow to infinity, apply **l'Hôpital's rule** to evaluate the limit. This rule states to take a derivative of each function with respect to x, and then evaluate each function at the limit.

$$\lim_{x \to c} \frac{f(x)}{g(x)} = \frac{\left.\frac{df}{dx}\right|_{x=c}}{\left.\frac{dg}{dx}\right|_{x=c}}$$

Example 1. $\lim_{x \to 1} \ln x = \ln 1 = 0$

Example 2. $\lim_{x \to 0^+} \frac{1}{x} = \infty$ since $\frac{1}{x}$ gets increasingly larger as x gets smaller (for example, $\frac{1}{0.01} = 100$)

Example 3. $\lim_{x \to 0^-} \frac{1}{x} = \infty$ since $\frac{1}{x}$ gets more and more negative as x becomes less negative (for example, $\frac{1}{-0.01} = -100$)

Example 4. $\lim_{x \to 0} \frac{\sin x}{x} = \frac{\left.\frac{d}{dx} \sin x\right|_{x=0}}{\left.\frac{d}{dx} x\right|_{x=0}} = \frac{\cos x|_{x=0}}{1} = \frac{\cos 0}{1} = \frac{1}{1} = 1$

Exercise Set 9.1

Directions: Evaluate the indicated limit.

1) $\lim_{x \to 0^+} \ln x =$

2) $\lim_{x \to 0} e^x =$

3) $\lim_{x \to \infty} e^{-x} =$

4) $\lim_{x \to -\infty} 2^{-x} =$

5) $\lim_{x \to 0^+} \operatorname{csch} x =$

6) $\lim_{x \to 0^-} \operatorname{csch} x =$

7) $\lim_{x \to 0} \operatorname{sech}^{-1} x =$

8) $\lim_{x \to 1^+} \cosh^{-1} x =$

9) $\lim_{x \to 0} \frac{\sinh x}{x} =$

10) $\lim_{x \to \infty} \frac{\ln x}{x^2} =$

9.2 The Definition of Euler's Number

Euler's number (Sec. 1.6) is defined by the following limit. Precisely, Euler's number is defined such that the numerical value of e satisfies the following equation.

$$\lim_{u \to 0} \frac{e^u - 1}{u} = 1$$

You will be able to explore this numerically in the exercises for this section.

Why does $\lim_{u \to 0} \frac{e^u - 1}{u}$ equal 1? To see this, let $y = e^u - 1$. Add 1 to both sides to get $y + 1 = e^u$. Take the natural logarithm of both sides to get $\ln(y + 1) = u$, according to the cancellation equation $\ln(e^u) = u$ from Chapter 3. Substitute $y = e^u - 1$ and $\ln(y + 1) = u$ into $\lim_{u \to 0} \frac{e^u - 1}{u}$, noting that $\lim_{u \to 0} y = \lim_{u \to 0} (e^u - 1) = e^0 - 1 = 1 - 1 = 0$ since $y = e^u - 1$:

$$\lim_{u \to 0} \frac{e^u - 1}{u} = \lim_{y \to 0} \frac{y}{\ln(y + 1)}$$

Since $\frac{1}{1/y} = 1 \div \frac{1}{y} = 1 \times \frac{y}{1} = y$ (since dividing by a fraction equates to multiplying by the fraction's reciprocal), we may rewrite the previous limit as:

$$\lim_{u \to 0} \frac{e^u - 1}{u} = \lim_{y \to 0} \frac{1}{\frac{1}{y} \ln(y + 1)}$$

Recall from Sec. 3.4 that $a \ln p = \ln(p^a)$, such that:

$$\lim_{u \to 0} \frac{e^u - 1}{u} = \lim_{y \to 0} \frac{1}{\ln(y + 1)^{1/y}} = \frac{1}{\ln \left[\lim_{y \to 0} (y + 1)^{1/y} \right]}$$

In Sec. 1.6, we learned that $e = \lim_{n \to \infty} \left(1 + \frac{1}{n}\right)^n$. If we let $y = \frac{1}{n}$, we get $e = \lim_{y \to 0} (y + 1)^{\frac{1}{y}}$. This allows us to replace $\ln \left[\lim_{y \to 0} (y + 1)^{1/y} \right]$ with $\ln e$ in the equation above.

$$\lim_{u \to 0} \frac{e^u - 1}{u} = \frac{1}{\ln e}$$

Since $\ln e = 1$, this becomes:

$$\lim_{u \to 0} \frac{e^u - 1}{u} = 1$$

This shows why $\lim_{u \to 0} \frac{e^u - 1}{u}$ is equal to one.

Example 1. Use a calculator to determine $\frac{e^u-1}{u}$ for $u = 0.1$.

$$\frac{e^{0.1}-1}{0.1} \approx \frac{1.10517-1}{0.1} \approx \frac{0.10517}{0.1} \approx 1.0517$$

Even though $u = 0.1$ is not particularly small, $\frac{e^u-1}{u}$ only differs from 1 by about 5% for this value of u. As you'll explore in the exercises that follow, the smaller u gets, the closer $\frac{e^u-1}{u}$ gets to 1.

Exercise Set 9.2

Directions: Use a calculator to determine each of the following.

1) $\frac{e^u-1}{u}$ for $u = 0.01$

2) $\frac{e^u-1}{u}$ for $u = 0.001$

3) $\frac{e^u-1}{u}$ for $u = 0.0001$

4) $\frac{e^u-1}{u}$ for $u = 0.00001$

9.3 Derivatives

The derivative of a function $f(x)$ evaluated at a is:

$$\left.\frac{df}{dx}\right|_{x=a} = \lim_{h \to 0} \frac{f(a+h) - f(a)}{h}$$

Example 1. Apply the limit definition of a derivative to derive an expression for the derivative of the exponential function, $y = e^{kx}$.

$$\frac{d}{dx} e^{kx} = \lim_{h \to 0} \frac{e^{k(x+h)} - e^{kx}}{h} = \lim_{h \to 0} \frac{e^{kx+kh} - e^{kx}}{h} = \lim_{h \to 0} \frac{e^{kx} e^{kh} - e^{kx}}{h}$$

$$= e^{kx} \lim_{h \to 0} \frac{e^{kh} - 1}{h} = e^{kx} \lim_{u \to 0} \frac{e^u - 1}{u/k} = \frac{e^{kx}}{1/k} \lim_{u \to 0} \frac{e^u - 1}{u} = k e^{kx}$$

Here are a few notes regarding the above steps:

- $e^{kx+kh} = e^{kx} e^{kh}$ because $x^{m+n} = x^m x^n$ (Sec. 1.5).
- We factored e^{kx} out to write: $e^{kx} e^{kh} - e^{kx} = e^{kx}(e^{kh} - 1)$.
- We defined $u = kh$ such that $h = u/k$.
- To divide by a fraction, multiply by its reciprocal: $\frac{e^{kx}}{u/k} = e^{kx} \div \frac{u}{k} = e^{kx} \times \frac{k}{u} = \frac{k e^{kx}}{u}$.
- In the last step, we used the definition of Euler's number (Sec. 9.2): $\lim_{u \to 0} \frac{e^u - 1}{u} = 1$.

If we omit the in-between steps from this example, we can express this result as:

$$\frac{d}{dx} e^{kx} = k e^{kx}$$

9 Calculus

Example 2. Apply the result of Example 1 to derive an expression for the derivative of $y = \ln x$.

Let $y = \ln x$. Exponentiate both sides to get $e^y = x$. (Recall from Chapter 3 that $e^{\ln x} = x$.) Take a derivative of both sides of $e^y = x$ with respect to x:

$$\frac{d}{dx}e^y = \frac{d}{dx}x$$

Apply the result from Example 1 that $\frac{d}{dx}e^{kx} = ke^{kx}$. Here, $k = 1$, but note that $\frac{d}{dx}e^y$ has two different variables, x and y. Recall the chain rule, $\frac{df}{dx} = \frac{df}{du}\frac{du}{dx}$ (which we will explore in Sec. 9.4). Here, $f = e^y$ and $u = y$ such that $\frac{df}{dx} = \frac{df}{du}\frac{du}{dx}$ is $\frac{d}{dx}e^y = \frac{d}{dy}e^y \frac{dy}{dx}$. The first part of this is $\frac{d}{dy}e^y = e^y$. Finally, note that $\frac{d}{dx}x = 1$. The previous equation, $\frac{d}{dx}e^y = \frac{d}{dx}x$, becomes $\frac{d}{dy}e^y \frac{dy}{dx} = 1$ which simplifies to:

$$e^y \frac{dy}{dx} = 1$$

Divide both sides of the equation by e^y:

$$\frac{dy}{dx} = \frac{1}{e^y}$$

Recall from the beginning of this solution that $y = \ln x$ and $e^y = x$. We may replace y with $\ln x$ and replace e^y with x in the previous equation.

$$\frac{d}{dx}\ln x = \frac{1}{x}$$

Example 3. Apply the result of Example 1 to derive an expression for the derivative of $y = \sinh kx$.

Recall from Chapter 6 that $\sinh x = \frac{e^x - e^{-x}}{2}$. Note that $\frac{d}{dx}e^{-kx} = -ke^{-kx}$.

$$\frac{d}{dx}\sinh kx = \frac{d}{dx}\frac{e^{kx} - e^{-kx}}{2} = \frac{1}{2}\frac{d}{dx}e^{kx} - \frac{1}{2}\frac{d}{dx}e^{-kx}$$

$$= \frac{ke^{kx}}{2} - \frac{1}{2}(-k)e^{-kx} = \frac{ke^{kx}}{2} + \frac{ke^{-kx}}{2} = k\frac{e^{kx} + e^{-kx}}{2} = k\cosh kx$$

Example 4. Apply the result of Example 2 to derive an expression for the derivative of $y = \sinh^{-1} x$.

Recall from Sec. 6.7 that $\sinh^{-1} x = \ln(x + \sqrt{x^2 + 1})$.

$$\frac{d}{dx} \sinh^{-1} x = \frac{d}{dx} \ln\left(x + \sqrt{x^2 + 1}\right)$$

Apply the chain rule, $\frac{df}{dx} = \frac{df}{du}\frac{du}{dx}$ (which we will explore in Sec. 9.4). Here, $f = \ln u$ and $u = x + \sqrt{x^2 + 1}$ such that $\frac{df}{dx} = \frac{df}{du}\frac{du}{dx}$ is $\frac{d}{dx} \ln u = \frac{d}{du} \ln u \frac{d}{dx}\left(x + \sqrt{x^2 + 1}\right)$.

$$\frac{d}{dx} \sinh^{-1} x = \frac{d}{du} \ln u \frac{d}{dx}\left(x + \sqrt{x^2 + 1}\right)$$

According to Example 2, $\frac{d}{du} \ln u = \frac{1}{u}$.

$$\frac{d}{dx} \sinh^{-1} x = \frac{1}{u} \frac{d}{dx}\left(x + \sqrt{x^2 + 1}\right)$$

Make the substitution $u = x + \sqrt{x^2 + 1}$.

$$\frac{d}{dx} \sinh^{-1} x = \frac{1}{x + \sqrt{x^2 + 1}} \frac{d}{dx}\left(x + \sqrt{x^2 + 1}\right)$$

Distribute the derivative according to the rule $\frac{d}{dx}(y_1 + y_2) = \frac{dy_1}{dx} + \frac{dy_2}{dx}$.

$$\frac{d}{dx} \sinh^{-1} x = \frac{1}{x + \sqrt{x^2 + 1}} \frac{d}{dx} x + \frac{1}{x + \sqrt{x^2 + 1}} \frac{d}{dx} \sqrt{x^2 + 1}$$

Note that $\frac{d}{dx} x = 1$.

$$\frac{d}{dx} \sinh^{-1} x = \frac{1}{x + \sqrt{x^2 + 1}} + \frac{1}{x + \sqrt{x^2 + 1}} \frac{d}{dx} \sqrt{x^2 + 1}$$

Use the chain rule again, this time with $u = x^2 + 1$ and $f = \sqrt{x^2 + 1} = \sqrt{u} = u^{1/2}$, to write: $\frac{d}{dx}\sqrt{x^2 + 1} = \frac{df}{dx} = \frac{df}{du}\frac{du}{dx} = \frac{d}{du} u^{1/2} \frac{d}{dx}(x^2 + 1) = \frac{1}{2} u^{-1/2}(2x) = \frac{x}{u^{1/2}} = \frac{x}{\sqrt{u}} = \frac{x}{\sqrt{x^2+1}}$. We used the rule $\frac{d}{dx} x^n = n \frac{d}{dx} x^{n-1}$ to write $\frac{d}{du} u^{1/2} = \frac{1}{2} u^{-1/2}$ and $\frac{d}{dx}(x^2 + 1) = \frac{d}{dx} x^2 + \frac{d}{dx} 1 = 2x$. Recall from Chapter 1 that $u^{-1/2} = \frac{1}{u^{1/2}} = \frac{1}{\sqrt{u}}$. Substitute $\frac{d}{dx}\sqrt{x^2 + 1} = \frac{x}{\sqrt{x^2+1}}$ into the above equation.

$$\frac{d}{dx} \sinh^{-1} x = \frac{1}{x + \sqrt{x^2 + 1}} + \frac{1}{x + \sqrt{x^2 + 1}} \frac{x}{\sqrt{x^2 + 1}}$$

$$\frac{d}{dx}\sinh^{-1} x = \frac{1}{x+\sqrt{x^2+1}}\frac{\sqrt{x^2+1}}{\sqrt{x^2+1}} + \frac{1}{x+\sqrt{x^2+1}}\frac{x}{\sqrt{x^2+1}}$$

$$\frac{d}{dx}\sinh^{-1} x = \frac{\sqrt{x^2+1}}{(x+\sqrt{x^2+1})(\sqrt{x^2+1})} + \frac{x}{(x+\sqrt{x^2+1})(\sqrt{x^2+1})}$$

$$\frac{d}{dx}\sinh^{-1} x = \frac{\sqrt{x^2+1}+x}{(x+\sqrt{x^2+1})(\sqrt{x^2+1})}$$

Note that $\sqrt{x^2+1}+x$ cancels in the numerator and denominator.

$$\frac{d}{dx}\sinh^{-1} x = \frac{1}{\sqrt{x^2+1}}$$

Alternative solution: The previous derivative can be found starting with $y = \sinh^{-1} x$. Take the hyperbolic sine of both sides: $\sinh y = x$. Take a derivative of both sides with respect to x.

$$\frac{d}{dx}\sinh y = \frac{d}{dx}x$$

Recall the chain rule, $\frac{df}{dx} = \frac{df}{du}\frac{du}{dx}$ (which we will explore in Sec. 9.4). Here, $f = \sinh y$ and $u = y$ such that $\frac{df}{dx} = \frac{df}{du}\frac{du}{dx}$ is $\frac{d}{dx}\sinh y = \frac{d}{dy}\sinh y \frac{dy}{dx}$. The first part of this is $\frac{d}{dy}\sinh y = \cosh y$. Finally, note that $\frac{d}{dx}x = 1$. The previous equation, $\frac{d}{dx}\sinh y = \frac{d}{dx}x$, becomes $\frac{d}{dy}\sinh y \frac{dy}{dx} = 1$ which simplifies to:

$$\cosh y \frac{dy}{dx} = 1$$

Divide by hyperbolic cosine on both sides.

$$\frac{dy}{dx} = \frac{1}{\cosh y}$$

Recall from Sec. 6.2, Problem 8, that $\cosh^2 y - \sinh^2 y = 1$. Add $\sinh^2 y$ to both sides to get $\cosh^2 y = 1 + \sinh^2 y$ and square root both sides: $\cosh y = \sqrt{1+\sinh^2 y}$. (We only need the positive root because $\cosh y$ is nonnegative, as seen in Chapter 7.) Plug $\cosh y = \sqrt{1+\sinh^2 y}$ into the previous equation.

$$\frac{dy}{dx} = \frac{1}{\sqrt{1+\sinh^2 y}}$$

Logarithms and Exponentials Essential Skills Practice Workbook with Answers

Recall from the beginning of this alternate solution that $y = \sinh^{-1} x$ and $\sinh y = x$. Substitute these expressions into the previous equation.

$$\frac{d}{dx}\sinh^{-1} x = \frac{1}{\sqrt{1+x^2}}$$

Exercise Set 9.3

Directions: Derive an expression for each derivative.

1) $\frac{d}{dx}\cosh kx =$

2) $\frac{d}{dx}\tanh kx =$

3) $\frac{d}{dx}\text{sech}\, kx =$

4) $\frac{d}{dx} \operatorname{csch} kx =$

5) $\frac{d}{dx} \coth kx =$

6) $\frac{d}{dx} b^x =$

7) $\frac{d}{dx} \log_{10} x =$

Logarithms and Exponentials Essential Skills Practice Workbook with Answers

8) $\dfrac{d}{dx}\cosh^{-1} x =$

9) $\dfrac{d}{dx}\tanh^{-1} x =$

10) $\dfrac{d}{dx}\text{sech}^{-1} x =$

9.4 The Product, Quotient, and Chain Rules

The **product rule** applies when you need to take a derivative of a product of functions of the same argument:

$$\frac{d}{dx}f(x)g(x) = g\frac{df}{dx} + f\frac{dg}{dx}$$

The **quotient rule** applies when you need to take a derivative of a ratio of functions of the same argument:

$$\frac{d}{dx}\frac{f(x)}{g(x)} = \frac{g\frac{df}{dx} - f\frac{dg}{dx}}{g^2}$$

The **chain rule** applies when you need to take a derivative of a function of one variable with respect to a second variable:

$$\frac{d}{dx}f(u(x)) = \frac{df}{du}\frac{du}{dx}$$

That is, $f(u)$ is a function of one variable u and $u(x)$ is a function of a second variable x.

We will explore these rules in the examples and exercises.

Example 1. Derive an expression for $\frac{d}{dx}xe^x$.

Use the product rule with $f = x$ and $g = e^x$. Recall from Sec. 9.3 that $\frac{d}{dx}e^x = e^x$.

$$\frac{d}{dx}xe^x = e^x\frac{d}{dx}x + x\frac{d}{dx}e^x = e^x(1) + x(e^x) = e^x + xe^x = e^x(1+x) = e^x(x+1)$$

Example 2. Derive an expression for $\frac{d}{dx}\frac{\ln x}{x}$.

Use the quotient rule with $f = \ln x$ and $g = x$. Recall from Sec. 9.3 that $\frac{d}{dx}\ln x = \frac{1}{x}$.

$$\frac{d}{dx}\frac{\ln x}{x} = \frac{x\frac{d}{dx}\ln x - \ln x\frac{d}{dx}x}{x^2} = \frac{x\left(\frac{1}{x}\right) - (\ln x)(1)}{x^2} = \frac{1 - \ln x}{x^2}$$

Logarithms and Exponentials Essential Skills Practice Workbook with Answers

Example 3. Derive an expression for $\frac{d}{dx}\ln|\sinh x|$.

Use the chain rule with $u = \sinh x$ and $f = \ln u$. Recall from Sec. 9.3 that $\frac{d}{du}\ln u = \frac{1}{u}$ and that $\frac{d}{dx}\sinh x = \cosh x$. Recall from Chapter 6 that $\coth x = \frac{\cosh x}{\sinh x}$.

$$\frac{d}{dx}\ln|\sinh x| = \frac{df}{du}\frac{du}{dx} = \frac{d}{du}\ln u \frac{d}{dx}\sinh x = \frac{1}{u}\cosh x = \frac{1}{\sinh x}\cosh x = \coth x$$

The absolute values reflect that a logarithm is only real if the argument is positive.

Exercise Set 9.4

Directions: Apply the product rule, quotient rule, or chain rule.

1) $\frac{d}{dx}e^{(x^2)} =$

2) $\frac{d}{dx}x\ln x =$

3) $\frac{d}{dx}\frac{1+\ln x}{1-\ln x} =$

4) $\frac{d}{dx}\ln(kx) =$

5) $\frac{d}{dx}\text{sech}^2 x =$

6) $\frac{d}{dx}\sinh x \cosh x =$

7) $\frac{d}{dx}\tanh^{-1}(\cos x) =$

Note: The inverse tangent is hyperbolic, but the cosine isn't.

8) $\dfrac{d}{dx}\sqrt{\tanh x} =$

9) $\dfrac{d}{dx}\sinh(e^x) =$

10) $\dfrac{d}{dx}x^x =$

9.5 Series Expansions

A **sequence** refers to a list of numbers that has a definite order, whereas a **series** refers to a sequence of terms that are added together. The uppercase Greek letter sigma (Σ) is the standard notation for a series.

$$\sum_{n=1}^{25} \frac{1}{n^2} = \frac{1}{1^2} + \frac{1}{2^2} + \frac{1}{3^2} + \frac{1}{4^2} + \cdots + \frac{1}{25^2}$$

The series above begins with the lower limit ($n = 1$), increases the value of the index (n) in increments of one, and continues to the upper limit ($n = 25$).

A **power series** is a series that has the following form:

$$\sum_{n=0}^{\infty} c_n x^n = c_0 + c_1 x + c_2 x^2 + c_3 x^3 + \cdots$$

The constants c_0, c_1, c_2, c_3, etc. are the coefficients of the series. Each term raises the variable to a different power.

If a function $f(x)$ can be expanded in a **power series** at $x = a$, it will have the form of a **Taylor series** expansion:

$$f(x) = \sum_{n=0}^{\infty} \frac{f^{(n)}(a)(x-a)^n}{n!} = f(a) + \frac{f'(a)}{1!}(x-a) + \frac{f''(a)}{2!}(x-a)^2 + \cdots$$

As mentioned at the end of Chapter 1, the exclamation mark (!) indicates a factorial, which means to multiply successively smaller integers until reaching 1. For example, $4! = 4(3)(2)(1) = 24$. Also as discussed in Chapter 1, note that $1! = 1$ and $0! = 1$. The notation $f^{(n)}$ refers to the nth derivative of f with respect to x, and the notation $f'(a)$, $f''(a), f'''(a)$, etc. means to take one derivative for each apostrophe (') and evaluate the derivative at $x = a$. For example, $f''(a)$ is the second derivative evaluated at a.

It is common to expand a function about $x = 0$, which corresponds to setting $a = 0$ in the Taylor series. This special case is referred to as the **Maclaurin series** expansion.

$$f(x) = \sum_{n=0}^{\infty} \frac{f^{(n)}(0) x^n}{n!} = f(0) + \frac{f'(0)}{1!} x + \frac{f''(0)}{2!} x^2 + \frac{f'''(0)}{3!} x^3 + \cdots$$

For example, the Maclaurin series for the exponential function, $f(x) = e^x$, is:

$$e^x = e^0 + \frac{e^0}{1!}x + \frac{e^0}{2!}x^2 + \frac{e^0}{3!}x^3 + \cdots = 1 + x + \frac{x^2}{2} + \frac{x^3}{6} + \frac{x^4}{24} + \frac{x^5}{120} + \cdots = \sum_{n=0}^{\infty} \frac{x^n}{n!}$$

Note that $\frac{d}{dx}e^x = e^x$ and $e^0 = 1$, such that $f(0) = 1, f'(0) = 1, f''(0) = 1, f'''(0) = 1$, etc. For example, $f'(0) = \frac{d}{dx}e^x\Big|_{x=0} = e^x|_{x=0} = e^0 = 1$, where the notation $\frac{d}{dx}e^x\Big|_{x=0}$ means to first take a derivative of e^x with respect to x and then evaluate the result at $x = 0$. The series above is valid for all values of x.

The Maclaurin series expansion for e^x provides a means of calculating the numerical value of Euler's number. If we set $x = 1$, we get a series for Euler's constant:

$$e^1 = e = 1 + 1 + \frac{1}{2} + \frac{1}{6} + \frac{1}{24} + \frac{1}{120} + \cdots = \sum_{n=0}^{\infty} \frac{1}{n!}$$

Recall that we explored the above series at the end of Sec. 1.6.

Two common Maclaurin series involving natural logarithms include:

$$\ln(1-x) = -x - \frac{x^2}{2} - \frac{x^3}{3} - \cdots = -\sum_{n=0}^{\infty} \frac{x^n}{n} \quad \text{for} -1 \leq x < 1$$

$$\ln(1+x) = x - \frac{x^2}{2} + \frac{x^3}{3} - \cdots = \sum_{n=0}^{\infty} (-1)^{n+1} \frac{x^n}{n} \quad \text{for} -1 < x \leq 1$$

The factor $(-1)^{n+1}$ creates the alternating signs for the second series.

The Maclaurin series expansions for sine and cosine share a remarkable similarity with the Maclaurin series expansions for the exponential function. In Sec. 10.4, we will see that these functions are related in the context of complex numbers.

$$\sin x = \sin 0 + \frac{\cos 0}{1!}x - \frac{\sin 0}{2!}x^2 - \frac{\cos 0}{3!}x^3 + \frac{\sin 0}{4!}x^4 + \frac{\cos 0}{5!}x^5 - \frac{\sin 0}{6!}x^6 - \frac{\cos 0}{7!}x^7 + \cdots$$

$$\sin x = 0 + (1)x - 0 - (1)\frac{x^3}{3!} + 0 + (1)\frac{x^5}{5!} - 0 - (1)\frac{x^7}{7!} + \cdots$$

$$\sin x = x - \frac{x^3}{3!} + \frac{x^5}{5!} - \frac{x^7}{7!} + \cdots = x - \frac{x^3}{6} + \frac{x^5}{120} - \frac{x^7}{5040} + \cdots$$

$$\cos x = \cos 0 - \frac{\sin 0}{1!}x - \frac{\cos 0}{2!}x^2 + \frac{\sin 0}{3!}x^3 + \frac{\cos 0}{4!}x^4 - \frac{\sin 0}{5!}x^5 - \frac{\cos 0}{6!}x^6 + \frac{\sin 0}{7!}x^7 + \cdots$$

$$\cos x = 1 - 0 - (1)\frac{x^2}{2!} + 0 + (1)\frac{x^4}{4!} - 0 - (1)\frac{x^6}{6!} + 0 + \cdots$$

$$\cos x = 1 - \frac{x^2}{2!} + \frac{x^4}{4!} - \frac{x^6}{6!} + \cdots = 1 - \frac{x^2}{2} + \frac{x^4}{24} - \frac{x^6}{720} + \cdots$$

Note that x must be expressed in radians for the above series, where π rad $= 180°$. The Maclaurin series expansions for hyperbolic sine and hyperbolic cosine share a remarkable similarity with the Maclaurin series expansions for sine and cosine. In Sec. 10.8, we will see that these functions are related in the context of complex numbers. The similarity between the Maclaurin series expansions for $\sin x$ and $\sinh x$ and for $\cos x$ and $\cosh x$ stems from the fact that $\sin 0 = 0$ and $\sinh 0 = \frac{e^0 - e^{-0}}{2} = \frac{1-1}{2} = 0$ and that $\cos 0 = 1$ and $\cosh 0 = \frac{e^0 + e^{-0}}{2} = \frac{1+1}{2} = \frac{2}{2} = 1$, as well as the fact that $\frac{d}{dx}\sin x = \cos x$ and $\frac{d}{dx}\sinh x = \cosh x$ and that $\frac{d}{dx}\cos x = -\sin x$ and $\frac{d}{dx}\cosh x = \sinh x$. There is just one significant difference between the corresponding series: the series for $\sinh x$ and $\cosh x$ do not have any minus signs. The reason for this is that $\frac{d}{dx}\cos x = -\sin x$ has a minus sign, whereas $\frac{d}{dx}\cosh x = \sinh x$ does not (see the solution to Problem 1 in Sec. 9.3).

$$\sinh x = \sinh 0 + \frac{\cosh 0}{1!}x + \frac{\sinh 0}{2!}x^2 + \frac{\cosh 0}{3!}x^3 + \frac{\sinh 0}{4!}x^4 + \frac{\cosh 0}{5!}x^5 + \frac{\sinh 0}{6!}x^6 + \cdots$$

$$\sinh x = 0 + (1)x + 0 + (1)\frac{x^3}{3!} + 0 + (1)\frac{x^5}{5!} + 0 + (1)\frac{x^7}{7!} + \cdots$$

$$\sinh x = x + \frac{x^3}{3!} + \frac{x^5}{5!} + \frac{x^7}{7!} + \cdots = x + \frac{x^3}{6} + \frac{x^5}{120} + \frac{x^7}{5040} + \cdots$$

$$\cosh x = \cosh 0 - \frac{\sinh 0}{1!}x - \frac{\cosh 0}{2!}x^2 + \frac{\sinh 0}{3!}x^3 + \frac{\cosh 0}{4!}x^4 - \frac{\sinh 0}{5!}x^5 - \frac{\cosh 0}{6!}x^6 + \cdots$$

$$\cosh x = 1 + 0 + (1)\frac{x^2}{2!} + 0 + (1)\frac{x^4}{4!} + 0 + (1)\frac{x^6}{6!} + 0 + \cdots$$

$$\cosh x = 1 + \frac{x^2}{2!} + \frac{x^4}{4!} + \frac{x^6}{6!} + \cdots = 1 + \frac{x^2}{2} + \frac{x^4}{24} + \frac{x^6}{720} + \cdots$$

Logarithms and Exponentials Essential Skills Practice Workbook with Answers

Example 1. Use the Maclaurin series to determine e^3 up to $n = 6$.

$$e^3 \approx 1 + 3 + \frac{3^2}{2} + \frac{3^3}{6} + \frac{3^4}{24} + \frac{3^5}{120} + \frac{3^6}{720} = 19.4125$$

Compare this to $e^3 \approx 20.0855$. For a higher value of n, the approximation would be better.

Exercise Set 9.5

Directions: Use a Maclaurin series expansion to calculate each value up to the specified value of n. Compare with the actual value (as obtained with a calculator).

1) e^{-1} for $n = 7$

2) e^2 for $n = 7$

3) $\sin\left(\frac{\pi}{6}\right)$ up to 3 nonzero terms

4) $\ln 2$ for $n = 7$

9.6 Integrals

The indefinite integrals of the basic exponential, natural logarithm, and hyperbolic functions are:

$$\int e^{kx}\,dx = \frac{e^{kx}}{k} + c \quad, \quad \int b^x\,dx = \frac{b^x}{\ln b} + c \quad, \quad \int \ln x\,dx = x \ln x - x + c$$

$$\int \sinh x\,dx = \cosh x + c \quad, \quad \int \cosh x\,dx = \sinh x + c$$

$$\int \tanh x\,dx = \ln|\cosh x| + c \quad, \quad \int \coth x\,dx = \ln|\sinh x| + c$$

$$\int \operatorname{csch} x\,dx = \ln\left|\tanh\left(\frac{x}{2}\right)\right| + c \quad, \quad \int \operatorname{sech} x\,dx = \tan^{-1}|\sinh x| + c$$

Note that $\int \sinh x\,dx$ and $\int \cosh x\,dx$ are both positive, unlike the similar integrals for ordinary trig functions. See the note to the solution to Problem 1 in Sec. 9.3. You can check the above formulas by taking a derivative. For example:

$$\frac{d}{dx}(x \ln x - x + c) = \frac{d}{dx}(x \ln x) - \frac{d}{dx}x + \frac{d}{dx}c$$

$$= \ln x \frac{d}{dx}x + x \frac{d}{dx}\ln x - \frac{d}{dx}x + \frac{d}{dx}c$$

$$= (\ln x)(1) + x\left(\frac{1}{x}\right) - 1 + 0 = \ln x + 1 - 1 = \ln x$$

Note that the integral for hyperbolic secant involves an ordinary inverse tangent (it's not hyperbolic) and a hyperbolic sine.

Example 1. Use integration by parts to show that $\int \ln x\,dx = x \ln x - x + c$.

Recall the integration by parts formula from calculus.

$$\int u\,dv = uv - \int v\,du$$

Let $u = \ln x$ and $dv = dx$ such that $du = \frac{dx}{x}$ (since $\frac{d}{dx}\ln x = \frac{1}{x}$) and $v = x$.

$$\int \ln x\,dx = (\ln x)(x) - \int x \frac{dx}{x} = x \ln x - \int dx = x \ln x - x + c$$

Logarithms and Exponentials Essential Skills Practice Workbook with Answers

Example 2. Use a substitution to show that $\int \tanh x \, dx = \ln|\cosh x| + c$.

Recall from Chapter 6 that $\tanh x = \frac{\sinh x}{\cosh x}$.

$$\int \tanh x \, dx = \int \frac{\sinh x}{\cosh x} dx$$

Let $u = \cosh x$ such that $du = \sinh x \, dx$ (since $\frac{d}{dx} \cosh x = \sinh x$).

$$\int \tanh x \, dx = \int \frac{du}{u} = \ln u + c = \ln|\cosh x| + c$$

The absolute values reflect that a logarithm is only real if the argument is positive.

Exercise Set 9.6

Directions: Perform each indefinite integral.

1) $\int 2^x \, dx =$

2) $\int \coth x \, dx =$

3) $\int \operatorname{sech} x \, dx =$

4) $\int \operatorname{csch} x \, dx =$

5) $\int x \sinh x \, dx =$

6) $\int x e^x \, dx =$

7) $\int \frac{\ln x}{x^2} \, dx =$

10 Complex Numbers

10.1 Real and Imaginary Parts

A **complex number** has the following form:
$$z = x + iy$$

- z represents the complex number; it has real and imaginary parts.
- x and y are real numbers; x represents the real part and y represents the imaginary part of the complex number.
- i is the **imaginary number**; it has the property that $i^2 = -1$.

No real number squared can ever be negative. An imaginary number is a number that when squared results in a negative value. For example, $(3i)^2 = 3^2 i^2 = 9(-1) = -9$.

For example, consider the complex number $2 - 3i$. The real part is 2 and the imaginary part is -3 (with the imaginary number, i, representing the imaginary part).

To add or subtract two complex numbers, add or subtract their real and imaginary parts. For example, $(5 + 8i) + (4 + 6i) = (5 + 4) + (8 + 6)i = 9 + 14i$. To multiply two complex numbers, apply the FOIL method from algebra and note that $i^2 = -1$. For example, $(4 + 3i)(2 + 5i) = 4(2) + 4(5i) + 3i(2) + 3i(5i) = 8 + 20i + 6i + 15i^2$
$= 8 + 26i + 15(-1) = 8 + 26i - 15 = -7 + 26i$.

The **complex conjugate** of a complex number, denoted by an asterisk (*), is found by negating the imaginary part.
$$z^* = x - iy$$
For example, for the complex number $2 + 7i$, its complex conjugate is $2 - 7i$.

When a complex number is multiplied by its complex conjugate, the result is the sum of the squares of the real and imaginary parts:
$$zz^* = z^*z = (x + iy)(x - iy) = x^2 - ixy + ixy - i^2 y^2 = x^2 + y^2$$

10 Complex Numbers

As the previous equation shows, the product of any complex number and its complex conjugate is always real (since $x^2 + y^2$ is a real number; there is no i in this expression).

The sum of the squares of the real and imaginary parts is referred to as the **modulus squared** and is denoted $|z|^2$.

$$|z|^2 = x^2 + y^2 = zz^* = z^*z$$

The positive square root of this is referred to as the **modulus**.

$$|z| = \sqrt{|z|^2} = \sqrt{x^2 + y^2} = \sqrt{zz^*} = \sqrt{z^*z}$$

If a fraction has a denominator that is a complex number, it is conventional to multiply the numerator and denominator each by the complex conjugate of the denominator. This makes the new denominator a real number. For example:

$$\frac{4-i}{3+2i} = \frac{4-i}{3+2i}\frac{3-2i}{3-2i} = \frac{12-8i-3i+2i^2}{9+4} = \frac{12-11i+2(-1)}{13} = \frac{10-11i}{13}$$

(This may remind you of the convention of rationalizing the denominator when the denominator of a fraction involves a square root. For example, $\frac{1}{\sqrt{2}} = \frac{1}{\sqrt{2}}\frac{\sqrt{2}}{\sqrt{2}} = \frac{\sqrt{2}}{2}$.)

The powers of i form a repeating pattern: $i^1 = i$, $i^2 = -1$, $i^3 = -i$, $i^4 = 1$, $i^5 = i$, $i^6 = -1$, $i^7 = -i$, $i^8 = 1$, $i^9 = i$, etc. Note that $(i^4)^n = 1$ if n is an integer.

Example 1. $(8 + 2i)(5 - 3i) = 8(5) + 8(-3i) + 2i(5) + 2i(-3i)$
$= 40 - 24i + 10i - 6i^2 = 40 - 14i - 6(-1) = 46 - 14i$

Example 2. $(5 + 4i)(5 + 4i)^* = (5 + 4i)(5 - 4i) = 5^2 + 4^2 = 25 + 16 = 41$
Note: $(5 + 4i)(5 - 4i) = 5(5) + 5(-4i) + 4i(5) + 4i(-4i) = 25 - 16(-1) = 25 + 16 = 41$. It's simpler to use the formula $(x + iy)(x - iy) = x^2 + y^2$.

Example 3. $\frac{2+i}{3-4i} = \frac{2+i}{3-4i}\frac{3+4i}{3+4i} = \frac{2(3)+2(4i)+i(3)+i(4i)}{3^2+(-4)^2} = \frac{6+8i+3i+4i^2}{9+16} = \frac{6+11i+4(-1)}{25} = \frac{2+11i}{25}$

Example 4. $i^{23} = i^{20}i^3 = (i^4)^5 i^3 = 1^5 i^3 = 1(-i) = -i$
Main idea: Find the largest power smaller than 23 that is divisible by 4 since $i^4 = 1$.

Logarithms and Exponentials Essential Skills Practice Workbook with Answers

Exercise Set 10.1

Directions: Express each answer in the form $a + bi$.

1) $(9 + 6i)(7 - 8i) =$

2) $(7 - 9i)^2 =$

3) $(1 - i)^3 =$

4) $(4 + 9i)(4 + 9i)^* =$

5) $(7 - 6i)(7 - 6i)^* =$

10 Complex Numbers

Directions: Make an equivalent fraction with a real denominator.

6) $\dfrac{2+i}{4+5i} =$

7) $\dfrac{2-7i}{8-3i} =$

Directions: Simplify each expression.

8) $i^{14} =$

9) $i^{25} =$

10) $i^{2027} =$

11) $i^{1000} =$

12) $\dfrac{1}{i} =$

13) $i^{-11} =$

10.2 The Complex Plane

A complex number $z = x + iy$ can be represented graphically by forming an ordered pair (x, y) from the real and imaginary parts and plotting the point in the plane. This is referred to as the **complex plane** (also called the **Argand plane**).
- The x-axis (horizontal) serves as the real axis. A real number lies on the x-axis.
- The y-axis (vertical) serves as the imaginary axis. An imaginary number lies on the y-axis.
- A complex number having both real and imaginary parts lies between axes.

The distance of the point (x, y) from the origin equals the modulus (Sec. 10.1).
$$|z| = \sqrt{|z|^2} = \sqrt{x^2 + y^2} = \sqrt{zz^*} = \sqrt{z^*z}$$

Example 1. Graph the complex number $4 + 3i$ in the complex plane. How far is this point from the origin?

The real part is 4 and the imaginary part is 3. Plot the point $(4, 3)$ by going 4 units to the right and 3 units up.

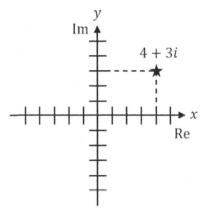

The point $(4, 3)$ plotted above is $\sqrt{4^2 + 3^2} = \sqrt{16 + 9} = \sqrt{25} = 5$ units from the origin.

Exercise Set 10.2

Directions: Graph each complex number in the complex plane and determine how far each point is from the origin.

1) $2 - 5i$

2) $4i$

3) $-4 - 4i$

4) $-3 + 2i$

10.3 Polar Form of Complex Numbers

A complex number $x + iy$ can be plotted in the complex plane as the ordered pair (x, y) with the real part (x) on the horizontal (real) axis and the imaginary part (y) on the vertical (imaginary) axis.

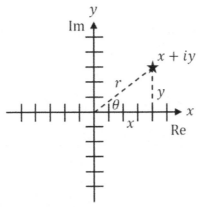

Just as any point (x, y) in the Cartesian plane can be expressed in 2D polar coordinates, a complex number $x + iy$ can be expressed in a polar form (r, θ), where, as usual, the polar coordinates are related to the horizontal and vertical coordinates by:

$$x = r \cos \theta \quad , \quad y = r \sin \theta$$

$$r = \sqrt{x^2 + y^2} = |z| = \sqrt{z^*z} = \sqrt{zz^*}$$

$$\theta = \tan^{-1}\left(\frac{y}{x}\right)$$

The polar coordinates of the complex number are interpreted as follows:
- r is the modulus (Sec. 10.1) of the complex number; it represents the distance of the point from the origin.
- θ is the angle counterclockwise from the real ($+x$) axis to r; it is conventional to express θ in radians, where π rad $= 180°$. Note that θ may exceed 2π rad (which is 360°). If you use the inverse tangent to find θ with a calculator, note that the calculator may not give you the answer in the quadrant that you need; you can reason out the correct quadrant from the signs of x and y (the same issue occurs in a standard trigonometry course in the context of real numbers).

Tip: When using a calculator, add π rad to the calculator's answer for $\tan^{-1}\left(\frac{y}{x}\right)$ if x is negative. (In that case, the answer lies in Quadrant II or III, but the calculator's inverse tangent is in Quadrant I or IV, which is opposite to the needed value.)

10 Complex Numbers

If we insert the previous equations for x and y into the equation $z = x + iy$, we get:
$$z = x + iy = r(\cos\theta + i\sin\theta)$$
The angle θ is called the argument of z. If you add or subtract an integral number of 2π's to θ, you get an alternate angle that is equivalent to θ in the sense that the new angle will give the same value for z.

Example 1. Write the complex number $-1 + i$ in polar form.

Compare $-1 + i$ with $z = x + iy$ to see that $x = -1$ and $y = 1$. First find r and θ.
$$r = |z| = \sqrt{x^2 + y^2} = \sqrt{(-1)^2 + 1^2} = \sqrt{1 + 1} = \sqrt{2}$$
$$\theta = \tan^{-1}\left(\frac{y}{x}\right) = \tan^{-1}\left(\frac{1}{-1}\right) = \tan^{-1}(-1) = \frac{3\pi}{4} \text{ rad}$$
$$z = r(\cos\theta + i\sin\theta) = \sqrt{2}\left[\cos\left(\frac{3\pi}{4}\right) + i\sin\left(\frac{3\pi}{4}\right)\right]$$

You can check the answer by applying trig. That is, since $\cos\left(\frac{3\pi}{4}\right) = -\frac{\sqrt{2}}{2}$ and $\sin\left(\frac{3\pi}{4}\right) = \frac{\sqrt{2}}{2}$, the above expression becomes $z = \sqrt{2}\left(-\frac{\sqrt{2}}{2} + i\frac{\sqrt{2}}{2}\right) = -\frac{\sqrt{2}\sqrt{2}}{2} + i\frac{\sqrt{2}\sqrt{2}}{2} = -1 + i$.
Note that the angle lies in Quadrant II. Since $x = -1$ is negative, add π to the calculator's answer to get the Quadrant II angle: $-\frac{\pi}{4} + \pi = \frac{3\pi}{4}$. (See the tip on the previous page.)

Example 2. If a complex number has $r = 4$ and $\theta = \frac{\pi}{6}$, write the complex number in the form $z = x + iy$.

First determine x and y.
$$x = r\cos\theta = 4\cos\left(\frac{\pi}{6}\right) = 4\left(\frac{\sqrt{3}}{2}\right) = 2\sqrt{3}$$
$$y = r\sin\theta = 4\sin\left(\frac{\pi}{6}\right) = 4\left(\frac{1}{2}\right) = 2$$
$$z = x + iy = 2\sqrt{3} + 2i$$

Logarithms and Exponentials Essential Skills Practice Workbook with Answers

Exercise Set 10.3

Directions: Determine r and θ for each complex number.

1) $6\sqrt{2} - 6i\sqrt{2}$

2) $-\frac{1}{4} - \frac{\sqrt{3}}{4}i$

3) $-5i$

4) -9

10.4 Euler's Formula

In Sec. 9.5, we saw that the exponential function e^x can be expanded in the following series. (Although we used calculus in Chapter 9, there isn't any calculus in Chapter 10.)

$$e^x = 1 + x + \frac{x^2}{2} + \frac{x^3}{6} + \frac{x^4}{24} + \frac{x^5}{120} + \cdots = \sum_{n=0}^{\infty} \frac{x^n}{n!}$$

This series applies even if the exponent is a complex number $z = x + iy$.

$$e^z = 1 + z + \frac{z^2}{2} + \frac{z^3}{6} + \frac{z^4}{24} + \frac{z^5}{120} + \cdots = \sum_{n=0}^{\infty} \frac{z^n}{n!}$$

Consider the special case where $z = i\theta$.

$$e^{i\theta} = 1 + i\theta + \frac{(i\theta)^2}{2} + \frac{(i\theta)^3}{6} + \frac{(i\theta)^4}{24} + \frac{(i\theta)^5}{120} + \cdots = \sum_{n=0}^{\infty} \frac{(i\theta)^n}{n!}$$

Recall that $i^1 = i, i^2 = -1, i^3 = -i, i^4 = 1, i^5 = i, i^6 = -1, i^7 = -i, i^8 = 1, i^9 = i$, etc.

$$e^{i\theta} = 1 + i\theta - \frac{\theta^2}{2} - i\frac{\theta^3}{6} + \frac{\theta^4}{24} + i\frac{\theta^5}{120} - \cdots$$

Every other term can be organized into two separate series as follows:

$$e^{i\theta} = \left(1 - \frac{\theta^2}{2} + \frac{\theta^4}{24} - \cdots\right) + i\left(\theta - \frac{\theta^3}{6} + \frac{\theta^5}{120} - \cdots\right)$$

The first series is the Maclaurin series expansion for cosine, while the second series is the Maclaurin series expansion for sine. (We discussed these series in Sec. 9.5.) Note that θ must be expressed in **radians** in these equations.

$$e^{i\theta} = \cos\theta + i\sin\theta$$

The above equation is known as **Euler's formula**. We can use Euler's number to express the polar form of a complex number as:

$$z = r(\cos\theta + i\sin\theta) = re^{i\theta}$$

Combining the two forms of a complex number (from Sec.'s 10.1 and 10.3), we get:

$$z = x + iy = re^{i\theta}$$

If we multiply Euler's formula by e^x (and change θ to y), we get:

$$e^x e^{iy} = e^{x+iy} = e^z = e^x(\cos y + i\sin y)$$

Example 1. $e^{i\pi} = \cos\pi + i\sin\pi = -1$ (Hence, the famous equation $e^{i\pi} + 1 = 0$, which uses the five common constants: 0, 1, e, π, and i.)

Example 2. $e^{1+i\pi/4} = e^1 e^{i\pi/4} = e^1\left[\cos\left(\frac{\pi}{4}\right) + i\sin\left(\frac{\pi}{4}\right)\right] = e\left(\frac{\sqrt{2}}{2} + i\frac{\sqrt{2}}{2}\right) = \frac{e\sqrt{2}}{2} + \frac{e\sqrt{2}}{2}i$

Example 3. Write $2 + 2i$ in the form $re^{i\theta}$.

Compare $2 + 2i$ with $z = x + iy$ to see that $x = 2$ and $y = 2$. First find r and θ.
$$r = |z| = \sqrt{x^2 + y^2} = \sqrt{2^2 + 2^2} = \sqrt{4+4} = \sqrt{8} = \sqrt{(4)(2)} = \sqrt{4}\sqrt{2} = 2\sqrt{2}$$
$$\theta = \tan^{-1}\left(\frac{y}{x}\right) = \tan^{-1}\left(\frac{2}{2}\right) = \tan^{-1}(1) = \frac{\pi}{4} \text{ rad}$$
$$z = re^{i\theta} = 2\sqrt{2}e^{i\pi/4}$$

Since $x = 2$ is positive, a calculator's answer would be correct for this problem. Since x and y are both positive, the angle lies in Quadrant I.

Exercise Set 10.4

Directions: Express each number in the form $x + iy$.

1) $e^{i\pi/3} =$

2) $e^{3i\pi/4} =$

3) $e^{5i\pi/6} =$

4) $e^{-1+3i\pi/2} =$

5) $e^{2+i\pi} =$

6) $e^{1-4i\pi/3} =$

Directions: Express each number in the form $re^{i\theta}$.

7) $\sqrt{3} - i =$

8) $-7i =$

10.5 De Moivre's Theorem

An efficient way to find the nth power of a complex number, for the case where n is a positive integer, is to apply **De Moivre's theorem**. According to De Moivre's theorem, for any positive integer n and a complex number expressed in polar form:
$$z^n = [r(\cos\theta + i\sin\theta)]^n = r^n[\cos(n\theta) + i\sin(n\theta)]$$
To raise a complex number to the power of a positive integer, raise the modulus ($r = |z|$) to the nth power and multiply the argument (θ) by n.

One way to see why De Moivre's theorem is true is to use Euler's formula (Sec. 10.4): $z = re^{i\theta} = r(\cos\theta + i\sin\theta)$. When we raise each side of $z = re^{i\theta}$ to the power of n, we get:
$$z^n = \left(re^{i\theta}\right)^n = (r)^n\left(e^{i\theta}\right)^n = r^n e^{in\theta}$$
Recall from Chapter 1 that $(x^m)^n = x^{mn}$ and that $(xy)^n = x^n y^n$. According to Euler's formula, $e^{iy} = \cos y + i\sin y$. If we let $y = n\theta$ in Euler's formula, we get $e^{in\theta} = \cos(n\theta) + i\sin(n\theta)$. If we substitute this into the above equation, we get De Moivre's theorem:
$$z^n = r^n e^{in\theta} = r^n[\cos(n\theta) + i\sin(n\theta)]$$

Example 1. Compute $(4 - 4i)^5$.

Compare $4 - 4i$ with $z = x + iy$ to see that $x = 4$ and $y = -4$. First find r and θ.
$$r = |z| = \sqrt{x^2 + y^2} = \sqrt{4^2 + (-4)^2} = \sqrt{16 + 16} = \sqrt{32} = \sqrt{(16)(2)} = \sqrt{16}\sqrt{2} = 4\sqrt{2}$$
$$\theta = \tan^{-1}\left(\frac{y}{x}\right) = \tan^{-1}\left(\frac{-4}{4}\right) = \tan^{-1}(-1) = -\frac{\pi}{4} \text{ rad}$$
$$z^n = r^n[\cos(n\theta) + i\sin(n\theta)] = (4 - 4i)^5 = \left(4\sqrt{2}\right)^5\left[\cos\left(-\frac{5\pi}{4}\right) + i\sin\left(-\frac{5\pi}{4}\right)\right]$$
$$(4 - 4i)^5 = 4^5\left(\sqrt{2}\right)^5\left(-\frac{\sqrt{2}}{2} + \frac{\sqrt{2}}{2}i\right) = 1024\left(\sqrt{2}\right)^5\frac{\sqrt{2}}{2}(-1 + i)$$
$$(4 - 4i)^5 = 512\left(\sqrt{2}\right)^6(-1 + i) = 512(8)(-1 + i) = -4096 + 4096i$$
Since $x = 4$ is positive, a calculator's answer would be correct for this problem. Since x is positive while y is negative, $-\frac{\pi}{4}$ rad lies in Quadrant IV. Note that $-\frac{\pi}{4}$ rad is equivalent to $\frac{7\pi}{4}$ rad since these angles differ by 2π. Also note that $n = 5$ and $n\theta = -\frac{5\pi}{4}$.

145

Exercise Set 10.5

Directions: Apply De Moivre's theorem to compute each of the following.

1) $(3 + 3i)^6 =$

2) $(\sqrt{3} + i)^{12} =$

3) $\left(-\frac{1}{2} + i\frac{\sqrt{3}}{2}\right)^3 =$

4) $(\sqrt{2} - i\sqrt{2})^7 =$

10.6 Roots of a Complex Number

Suppose that two complex numbers, z and u, are related by $z = u^n$, where n is a positive integer. In polar form, this relation is:
$$r(\cos\theta + i\sin\theta) = [v(\cos\varphi + i\sin\varphi)]^n$$
where $z = r(\cos\theta + i\sin\theta)$ and $u = v(\cos\varphi + i\sin\varphi)$. Note that $v = |u|$ is the modulus of u just like $r = |z|$ is the modulus of z and that φ is the argument of u just like θ is the argument of z. Apply De Moivre's theorem (Sec. 10.5) to the right-hand side.
$$r(\cos\theta + i\sin\theta) = v^n[\cos(n\varphi) + i\sin(n\varphi)]$$
The two sides of this equation are equal if the following are true:
$$r = v^n$$
$$\theta + 2\pi k = n\varphi$$
where k is an integer smaller than n. For example, if $n = 5$, then k may be 0, 1, 2, 3, or 4. The reason for the $2\pi k$ is that adding 2π (or 4π, 6π, etc.) to θ makes an equivalent angle. That is, $\cos(\theta + 2\pi k) = \cos\theta$ and $\sin(\theta + 2\pi k) = \sin\theta$. If we invert the above equations, we get:
$$v = r^{1/n}$$
$$\varphi = \frac{\theta + 2\pi k}{n}$$
This means that the nth root of a complex number $z = r(\cos\theta + i\sin\theta)$ can be found from the following formula, where n is a positive integer:
$$u = r^{1/n}\left[\cos\left(\frac{\theta + 2\pi k}{n}\right) + i\sin\left(\frac{\theta + 2\pi k}{n}\right)\right]$$
In fact, there are n different roots: one root for each value of k from 0 to $n-1$. Every root has the same modulus ($v = |u| = r^{1/n}$), but the argument $\left(\varphi = \frac{\theta + 2\pi k}{n}\right)$ is different for each root. If we plot all n roots in the complex plane (Sec. 10.2), they will all lie on a circle of radius $r^{1/n}$ centered about the origin, and the angle between two consecutive roots will equal $\frac{2\pi}{n}$ (such that the roots are evenly spaced in terms of angular separation). Such a graph is shown at the end of the example that follows.

10 Complex Numbers

Example 1. Find all of the cube roots of i.

Compare i with $z = x + iy$ to see that $x = 0$ and $y = 1$. First find r and θ.

$$r = |z| = \sqrt{x^2 + y^2} = \sqrt{0^2 + 1^2} = \sqrt{0 + 1} = \sqrt{1} = 1$$

$$\theta = \tan^{-1}\left(\frac{y}{x}\right) = \tan^{-1}\left(\frac{1}{0}\right) = \frac{\pi}{2} \text{ rad}$$

Note: Although $\frac{y}{x} = \frac{1}{0}$ is undefined, the angle is a definite $\frac{\pi}{2}$ rad. We can determine this geometrically because $x = 0$ and $y = 1$, corresponding to the $+y$-axis (at 90°). Since we want the cube roots, $n = 3$.

$$u = r^{1/n}\left[\cos\left(\frac{\theta + 2\pi k}{n}\right) + i\sin\left(\frac{\theta + 2\pi k}{n}\right)\right]$$

$$u = 1^{1/3}\left[\cos\left(\frac{\pi/2 + 2\pi k}{3}\right) + i\sin\left(\frac{\pi/2 + 2\pi k}{3}\right)\right]$$

$$u = 1\left[\cos\left(\frac{\pi}{6} + \frac{2\pi k}{3}\right) + i\sin\left(\frac{\pi}{6} + \frac{2\pi k}{3}\right)\right] = \cos\left(\frac{\pi}{6} + \frac{2\pi k}{3}\right) + i\sin\left(\frac{\pi}{6} + \frac{2\pi k}{3}\right)$$

For $k = 0$: $\quad u = \cos\left(\frac{\pi}{6}\right) + i\sin\left(\frac{\pi}{6}\right) = \frac{\sqrt{3}}{2} + \frac{i}{2}$

For $k = 1$: $\quad u = \cos\left(\frac{\pi}{6} + \frac{2\pi}{3}\right) + i\sin\left(\frac{\pi}{6} + \frac{2\pi}{3}\right) = \cos\left(\frac{5\pi}{6}\right) + i\sin\left(\frac{5\pi}{6}\right) = -\frac{\sqrt{3}}{2} + \frac{i}{2}$

For $k = 2$: $\quad u = \cos\left(\frac{\pi}{6} + \frac{4\pi}{3}\right) + i\sin\left(\frac{\pi}{6} + \frac{4\pi}{3}\right) = \cos\left(\frac{3\pi}{2}\right) + i\sin\left(\frac{3\pi}{2}\right) = 0 - i = -i$

The three cube roots i are $\frac{\sqrt{3}}{2} + \frac{i}{2}, -\frac{\sqrt{3}}{2} + \frac{i}{2}$, and $-i$. You can check the answers by cubing each complex number:

$$\left(\frac{\sqrt{3}}{2} + \frac{i}{2}\right)^3 = \left(\frac{\sqrt{3}}{2} + \frac{i}{2}\right)\left(\frac{\sqrt{3}}{2} + \frac{i}{2}\right)\left(\frac{\sqrt{3}}{2} + \frac{i}{2}\right) = \left(\frac{\sqrt{3}}{2} + \frac{i}{2}\right)\left(\frac{3}{4} + \frac{i\sqrt{3}}{2} - \frac{1}{4}\right)$$

$$= \left(\frac{\sqrt{3}}{2} + \frac{i}{2}\right)\left(\frac{1}{2} + \frac{i\sqrt{3}}{2}\right) = \frac{\sqrt{3}}{4} + \frac{3i}{4} + \frac{i}{4} - \frac{\sqrt{3}}{4} = i$$

$$\left(-\frac{\sqrt{3}}{2} + \frac{i}{2}\right)^3 = \left(-\frac{\sqrt{3}}{2} + \frac{i}{2}\right)\left(-\frac{\sqrt{3}}{2} + \frac{i}{2}\right)\left(-\frac{\sqrt{3}}{2} + \frac{i}{2}\right) = \left(-\frac{\sqrt{3}}{2} + \frac{i}{2}\right)\left(\frac{3}{4} - \frac{i\sqrt{3}}{2} - \frac{1}{4}\right)$$

$$= \left(-\frac{\sqrt{3}}{2} + \frac{i}{2}\right)\left(\frac{1}{2} - \frac{i\sqrt{3}}{2}\right) = -\frac{\sqrt{3}}{4} + \frac{3i}{4} + \frac{i}{4} + \frac{\sqrt{3}}{4} = i$$

$$(-i)^3 = (-i)(-i)(-i) = (-i)(i^2) = (-i)(-1) = i$$

Logarithms and Exponentials Essential Skills Practice Workbook with Answers

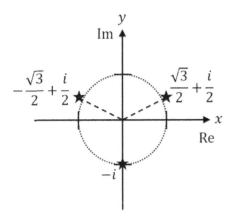

Exercise Set 10.6

Directions: Find all of the indicated roots of each number.

1) the fourth roots of $-1 - i\sqrt{3}$

2) the square roots of i

3) the cube roots of 27

Logarithms and Exponentials Essential Skills Practice Workbook with Answers

10.7 Roots of the Quadratic Equation

The quadratic formula from algebra has complex solutions when the **discriminant** ($b^2 - 4ac$) is negative. For example, consider the equation $x^2 + 5x + 8 = 0$. Compare this with the standard form $ax^2 + bx + c = 0$ to see that $a = 1$, $b = 5$, and $c = 8$. The discriminant is $b^2 - 4ac = 5^2 - 4(1)(8) = 25 - 32 = -7$. Since the discriminant is negative, the solutions to $x^2 + 5x + 8 = 0$ are complex:

$$x = \frac{-b \pm \sqrt{b^2 - 4ac}}{2a} = \frac{-5 \pm \sqrt{5^2 - 4(1)(8)}}{2(1)} = \frac{-5 \pm \sqrt{-7}}{2} = \frac{-5 \pm i\sqrt{7}}{2}$$

Note that $\sqrt{-7} = \sqrt{(-1)(7)} = \sqrt{-1}\sqrt{7} = i\sqrt{7}$. The two complex solutions, $x = -\frac{5}{2} + \frac{i\sqrt{7}}{2}$ and $x = -\frac{5}{2} - \frac{i\sqrt{7}}{2}$, are complex conjugates of one another (because they only differ in the sign of the imaginary part).

Example 1. Find all of the solutions to $x^2 + 3x + 5 = 0$.

Compare $x^2 + 3x + 5 = 0$ with the standard form $ax^2 + bx + c = 0$ to see that $a = 1$, $b = 3$, and $c = 5$. The discriminant is $b^2 - 4ac = 3^2 - 4(1)(5) = 9 - 20 = -11$. Since the discriminant is negative, the solutions to $x^2 + 3x + 5 = 0$ are complex:

$$x = \frac{-b \pm \sqrt{b^2 - 4ac}}{2a} = \frac{-3 \pm \sqrt{3^2 - 4(1)(5)}}{2(1)} = \frac{-3 \pm \sqrt{-11}}{2} = \frac{-3 \pm i\sqrt{11}}{2}$$

Note that $\sqrt{-11} = \sqrt{(-1)(11)} = \sqrt{-1}\sqrt{11} = i\sqrt{11}$.

Exercise Set 10.7

Directions: Find all of the solutions to each equation.

1) $2x^2 + 7x + 9 = 0$

2) $5x^2 - 10x + 6 = 0$

10.8 Relating Hyperbolic and Trig Functions

In Sec. 9.5, we saw that $\sin x$, $\cos x$, $\sinh x$, and $\cosh x$, can be expanded in the following series. (Although we used calculus in Chapter 9, there isn't any calculus in Chapter 10.)

$$\sin x = x - \frac{x^3}{3!} + \frac{x^5}{5!} - \frac{x^7}{7!} + \cdots \quad , \quad \sinh x = x + \frac{x^3}{3!} + \frac{x^5}{5!} + \frac{x^7}{7!} + \cdots$$

$$\cos x = 1 - \frac{x^2}{2!} + \frac{x^4}{4!} - \frac{x^6}{6!} + \cdots \quad , \quad \cosh x = 1 + \frac{x^2}{2!} + \frac{x^4}{4!} + \frac{x^6}{6!} + \cdots$$

Comparing these series expansions, we see that the hyperbolic sine and cosine are related to the ordinary sine and cosine via imaginary numbers. (Recall that $i^1 = i$, $i^2 = -1$, $i^3 = -i$, $i^4 = 1$, $i^5 = i$, $i^6 = -1$, $i^7 = -i$, $i^8 = 1$, $i^9 = i$, etc.)

$$\cosh(ix) = \cos x \quad , \quad \sinh(ix) = i \sin x$$
$$\cos(ix) = \cosh x \quad , \quad \sin(ix) = i \sinh x$$

The last relation, $\sin(ix) = i \sinh x$, is sometimes expressed as $\sinh x = -i \sin(ix)$. Note that these two relations are equivalent. If you multiply both sides of $\sin(ix) = i \sinh x$ by i, you get $i \sin(ix) = -\sinh x$, which is equivalent to $\sinh x = -i \sin(ix)$. At first glance, it may seem that $\sin(ix) = i \sinh x$ and $\sinh x = -i \sin(ix)$ are inconsistent, but we have just shown they are actually equivalent.

Using only real numbers, recall from trigonometry that the values of $\sin x$ and $\cos x$ oscillate between -1 and 1. With complex numbers, other values are possible. For example, $\cos\left(\frac{\pi}{3}i\right) = \cosh\left(\frac{\pi}{3}\right) = \frac{e^{\pi/3} + e^{-\pi/3}}{2} \approx 1.6$. Similarly, the inverse sine and cosine can only return a real value if the argument lies between -1 and 1, but using complex numbers, other arguments are possible. For example, $\cos^{-1}(1.6) \approx \frac{\pi}{3}i \approx 1.047i$ since $\cos\left(\frac{\pi}{3}i\right) = \cosh\left(\frac{\pi}{3}\right) \approx 1.6$.

We can use the trig identities $\cos(x+y) = \cos x \cos y - \sin x \sin y$ and $\sin(x+y) = \sin x \cos y + \cos x \sin y$ and the hyperbolic identities $\cosh(x+y) = \cosh x \cosh y + \sinh x \sinh y$ and $\sinh(x+y) = \sinh x \cosh y + \cosh x \sinh y$ (from Sec. 6.2) to find the sine, cosine, hyperbolic sine, or hyperbolic cosine of a complex number:

$$\cos(x + iy) = \cos x \cos(iy) - \sin x \sin(iy) = \cos x \cosh y - i \sin x \sinh y$$
$$\sin(x + iy) = \sin x \cos(iy) + \cos x \sin(iy) = \sin x \cosh y + i \cos x \sinh y$$
$$\cosh(x + iy) = \cosh x \cosh(iy) + \sinh x \sinh(iy) = \cosh x \cos y + i \sinh x \sin y$$
$$\sinh(x + iy) = \sinh x \cosh(iy) + \cosh x \sinh(iy) = \sinh x \cos y + i \cosh x \sin y$$

For tangent and hyperbolic tangent, note that:
$$\tanh(ix) = i \tan x \quad , \quad \tan(ix) = i \tanh x$$

As noted earlier with sine, $\tan(ix) = i \tanh x$ is equivalent to $\tanh x = -i \tan(ix)$. If you multiply both sides of $\tan(ix) = i \tanh x$ by i, you get $i \tan(ix) = -\tanh x$, which is equivalent to $\tanh x = -i \tan(ix)$.

Example 1. $\sin(i) = i \sinh(1) = i \frac{e^1 - e^{-1}}{2} \approx 1.175i$

Example 2. $\sinh\left(\frac{\pi}{4}i\right) = i \sin\left(\frac{\pi}{4}\right) = \frac{\sqrt{2}}{2}i \approx 0.7071i$

Example 3. $\cos\left(\frac{\pi}{2} + i\right) = \cos\left(\frac{\pi}{2}\right) \cosh 1 - i \sin\left(\frac{\pi}{2}\right) \sinh 1 = (0)\frac{e^1 + e^{-1}}{2} - i(1)\frac{e^1 - e^{-1}}{2}$
$\approx -1.175i$

Logarithms and Exponentials Essential Skills Practice Workbook with Answers

Exercise Set 10.8

Directions: Use the relations of this section to determine each answer. Use a calculator where necessary.

1) $\cos i \approx$

2) $\sin(i\pi) \approx$

3) $\cosh(\pi i) \approx$

4) $\sinh\left(\frac{\pi}{6}i\right) \approx$

5) $\tanh\left(\frac{\pi}{4}i\right) \approx$

6) $\cosh\left(\frac{5\pi}{6}i\right) \approx$

7) $\sin\left(\frac{\pi}{2}+i\right) \approx$

8) $\cos\left(\frac{\pi}{6}-i\right) \approx$

9) $\sinh(2+\pi i) \approx$

10) $\cosh\left(\frac{1-\pi i}{2}\right) \approx$

ANSWER KEY

Chapter 1 Review of Exponents

Exercise Set 1.1

1) $6^2 = 36$
2) $8^3 = 512$
3) $9^0 = 1$
4) $7^1 = 7$
5) $4^4 = 256$
6) $2^8 = 256$
7) $(-1)^{25} = -1$
8) $15^2 = 225$
9) $2^{10} = 1024$
10) $6^4 = 1296$
11) $(-3)^3 = -27$
12) $(-9)^2 = 81$
13) $(-2)^5 = -32$
14) $(-5)^4 = 625$
15) $18^1 = 18$
16) $50^3 = 125{,}000$
17) $(-30)^2 = 900$
18) $100^0 = 1$
19) $6^3 = 216$
20) $0^9 = 0$
21) $8^2 = 64$
22) $(-10)^3 = -1000$
23) $(-20)^4 = 160{,}000$
24) $3^5 = 243$

Exercise Set 1.2

1) $\sqrt{49} = \pm 7$ since $(\pm 7)^2 = 49$
2) $\sqrt[3]{216} = 6$ since $6^3 = 6 \times 6 \times 6 = 36 \times 6 = 216$
3) $\sqrt[4]{256} = \pm 4$ since $(\pm 4)^4 = 4 \times 4 \times 4 \times 4 = 16 \times 16 = 256$
4) $\sqrt[3]{-125} = -5$ since $(-5)^3 = (-5) \times (-5) \times (-5) = 25 \times (-5) = -125$
5) $\sqrt{81} = \pm 9$ since $(\pm 9)^2 = 81$
6) $\sqrt[3]{512} = 8$ since $8^3 = 8 \times 8 \times 8 = 64 \times 8 = 512$
7) $\sqrt[4]{625} = \pm 5$ since $(\pm 5)^4 = 5 \times 5 \times 5 \times 5 = 25 \times 25 = 625$
8) $\sqrt[3]{-8000} = -20$ since $(-20)^3 = (-20) \times (-20) \times (-20) = 400 \times (-20) = -8000$
9) $\sqrt[9]{512} = 2$ since $2^9 = 2 \times 2 \times 2 \times 2 \times 2 \times 2 \times 2 \times 2 \times 2 = 8 \times 8 \times 8 = 64 \times 8 = 512$
10) $\sqrt[6]{1{,}000{,}000{,}000{,}000} = \pm 100$ since $(\pm 100)^6$ has $2 \times 6 = 12$ zeros

Logarithms and Exponentials Essential Skills Practice Workbook with Answers

Exercise Set 1.3

1) $7^{-2} = \left(\frac{1}{7}\right)^2 = \frac{1^2}{7^2} = \frac{1 \times 1}{7 \times 7} = \frac{1}{49}$

2) $9^{-1} = \left(\frac{1}{9}\right)^1 = \frac{1^1}{9^1} = \frac{1}{9}$

3) $\left(\frac{13}{19}\right)^{-1} = \left(\frac{19}{13}\right)^1 = \frac{19^1}{13^1} = \frac{19}{13}$

4) $\left(\frac{5}{9}\right)^{-2} = \left(\frac{9}{5}\right)^2 = \frac{9^2}{5^2} = \frac{9 \times 9}{5 \times 5} = \frac{81}{25}$

5) $5^{-4} = \left(\frac{1}{5}\right)^4 = \frac{1^4}{5^4} = \frac{1 \times 1 \times 1 \times 1}{5 \times 5 \times 5 \times 5} = \frac{1 \times 1}{25 \times 25} = \frac{1}{625}$

6) $7^{-3} = \left(\frac{1}{7}\right)^3 = \frac{1^3}{7^3} = \frac{1 \times 1 \times 1}{7 \times 7 \times 7} = \frac{1 \times 1}{49 \times 7} = \frac{1}{343}$

7) $\left(\frac{1}{3}\right)^{-4} = \left(\frac{3}{1}\right)^4 = \frac{3^4}{1^4} = \frac{3 \times 3 \times 3 \times 3}{1 \times 1 \times 1 \times 1} = \frac{9 \times 9}{1 \times 1} = \frac{81}{1} = 81$

8) $\left(\frac{4}{5}\right)^{-3} = \left(\frac{5}{4}\right)^3 = \frac{5^3}{4^3} = \frac{5 \times 5 \times 5}{4 \times 4 \times 4} = \frac{25 \times 5}{16 \times 4} = \frac{125}{64}$

9) $(-9)^{-2} = \left(\frac{-1}{9}\right)^2 = \frac{(-1) \times (-1)}{9 \times 9} = \frac{1}{81}$

10) $(-1)^{-1} = \left(\frac{-1}{1}\right)^1 = \frac{(-1)^1}{1^1} = \frac{-1}{1} = -1$

11) $(-6)^{-3} = \left(\frac{-1}{6}\right)^3 = \frac{(-1)^3}{6^3} = \frac{(-1) \times (-1) \times (-1)}{6 \times 6 \times 6} = \frac{1 \times (-1)}{36 \times 6} = -\frac{1}{216}$

12) $(-5)^{-4} = \left(\frac{-1}{5}\right)^4 = \frac{(-1)^4}{5^4} = \frac{(-1) \times (-1) \times (-1) \times (-1)}{5 \times 5 \times 5 \times 5} = \frac{1 \times 1}{25 \times 25} = \frac{1}{625}$

13) $1^{-7} = \left(\frac{1}{1}\right)^7 = \frac{1^7}{1^7} = \frac{1}{1} = 1$

14) $10^{-3} = \left(\frac{1}{10}\right)^3 = \frac{1^3}{10^3} = \frac{1 \times 1 \times 1}{10 \times 10 \times 10} = \frac{1 \times 1}{100 \times 10} = \frac{1}{1000}$ (equivalent to 0.001)

15) $48^{-1} = \left(\frac{1}{48}\right)^1 = \frac{1^1}{48^1} = \frac{1}{48}$

16) $\left(\frac{5}{2}\right)^{-3} = \left(\frac{2}{5}\right)^3 = \frac{2^3}{5^3} = \frac{2 \times 2 \times 2}{5 \times 5 \times 5} = \frac{4 \times 2}{25 \times 5} = \frac{8}{125}$

17) $\left(-\frac{4}{9}\right)^{-2} = \left(\frac{-9}{4}\right)^2 = \frac{(-9) \times (-9)}{4 \times 4} = \frac{81}{16}$

18) $\left(-\frac{1}{3}\right)^{-5} = \left(\frac{-3}{1}\right)^5 = \frac{(-3)^5}{1^5} = \frac{(-3) \times (-3) \times (-3) \times (-3) \times (-3)}{1 \times 1 \times 1 \times 1 \times 1} = \frac{9 \times 9 \times (-3)}{1 \times 1} = \frac{81 \times (-3)}{1 \times 1} = -\frac{243}{1} = -243$

19) $8^{-3} = \left(\frac{1}{8}\right)^3 = \frac{1^3}{8^3} = \frac{1 \times 1 \times 1}{8 \times 8 \times 8} = \frac{1 \times 1}{64 \times 8} = \frac{1}{512}$

Answer Key

20) $3^{-4} = \left(\frac{1}{3}\right)^4 = \frac{1^4}{3^4} = \frac{1\times1\times1\times1}{3\times3\times3\times3} = \frac{1\times1}{9\times9} = \frac{1}{81}$

21) $(-2)^{-9} = \left(\frac{-1}{2}\right)^9 = \frac{-1}{2\times2\times2\times2\times2\times2\times2\times2\times2} = -\frac{1}{8\times8\times8} = -\frac{1}{64\times8} = -\frac{1}{512}$

22) $\left(\frac{11}{12}\right)^{-1} = \left(\frac{12}{11}\right)^1 = \frac{12^1}{11^1} = \frac{12}{11}$

Exercise Set 1.4

1) $27^{4/3} = \left(\sqrt[3]{27}\right)^4 = 3^4 = 81$

2) $256^{1/4} = \left(\sqrt[4]{256}\right)^1 = (\pm 4)^1 = \pm 4$

3) $36^{1/2} = \left(\sqrt{36}\right)^1 = (\pm 6)^1 = \pm 6$

4) $64^{5/3} = \left(\sqrt[3]{64}\right)^5 = 4^5 = 1024$

5) $1{,}000{,}000^{5/6} = \left(\sqrt[6]{1{,}000{,}000}\right)^5 = (\pm 10)^5 = \pm 100{,}000$

6) $32^{4/5} = \left(\sqrt[5]{32}\right)^4 = 2^4 = 16$

7) $(-243)^{3/5} = \left(\sqrt[5]{-243}\right)^3 = (-3)^3 = -27$

8) $(-216)^{2/3} = \left(\sqrt[3]{-216}\right)^2 = (-6)^2 = 36$

9) $125^{-1/3} = \left(\frac{1}{125}\right)^{1/3} = \left(\sqrt[3]{\frac{1}{125}}\right)^1 = \left(\frac{1}{5}\right)^1 = \frac{1}{5}$

10) $121^{-3/2} = \left(\frac{1}{121}\right)^{3/2} = \left(\sqrt{\frac{1}{121}}\right)^3 = \left(\pm\frac{1}{11}\right)^3 = \pm\frac{1}{1331}$

11) $\left(\frac{4}{9}\right)^{5/2} = \left(\sqrt{\frac{4}{9}}\right)^5 = \left(\pm\frac{2}{3}\right)^5 = \pm\frac{32}{243}$

12) $\left(\frac{625}{81}\right)^{3/4} = \left(\sqrt[4]{\frac{625}{81}}\right)^3 = \left(\pm\frac{5}{3}\right)^3 = \pm\frac{125}{27}$

13) $\left(\frac{32}{243}\right)^{1/5} = \left(\sqrt[5]{\frac{32}{243}}\right)^1 = \left(\frac{2}{3}\right)^1 = \frac{2}{3}$

14) $\left(\frac{1}{256}\right)^{-1/8} = (256)^{1/8} = \left(\sqrt[8]{256}\right)^1 = (\pm 2)^1 = \pm 2$

15) $\left(-\frac{8}{27}\right)^{5/3} = \left(\sqrt[3]{-\frac{8}{27}}\right)^5 = \left(-\frac{2}{3}\right)^5 = -\frac{32}{243}$

Logarithms and Exponentials Essential Skills Practice Workbook with Answers

16) $\left(-\frac{243}{1024}\right)^{2/5} = \left(\sqrt[5]{-\frac{243}{1024}}\right)^2 = \left(-\frac{3}{4}\right)^2 = \frac{9}{16}$

17) $160{,}000^{5/4} = \left(\sqrt[4]{160{,}000}\right)^5 = (\pm 20)^5 = \pm 3{,}200{,}000$

18) $(-1)^{-8/7} = \left(-\frac{1}{1}\right)^{8/7} = (-1)^{8/7} = \left(\sqrt[7]{-1}\right)^8 = (-1)^8 = 1$

19) $225^{1/2} = \left(\sqrt{225}\right)^1 = (\pm 15)^1 = \pm 15$

20) $256^{-3/4} = \left(\frac{1}{256}\right)^{3/4} = \left(\sqrt[4]{\frac{1}{256}}\right)^3 = \left(\pm\frac{1}{4}\right)^3 = \pm\frac{1}{64}$

21) $\left(-\frac{1}{100{,}000}\right)^{-9/5} = (-100{,}000)^{9/5} = \left(\sqrt[5]{-100{,}000}\right)^9 = (-10)^9 = -1{,}000{,}000{,}000$

22) $\left(-\frac{343}{512}\right)^{-2/3} = \left(-\frac{512}{343}\right)^{2/3} = \left(\sqrt[3]{-\frac{512}{343}}\right)^2 = \left(-\frac{8}{7}\right)^2 = \frac{64}{49}$

Exercise Set 1.5

1) $x^4 x^2 = x^{4+2} = x^6$

2) $x^8 x^{-4} = x^{8+(-4)} = x^4$

3) $x^5 x^4 x = x^5 x^4 x^1 = x^{5+4+1} = x^{10}$

4) $x^2 x^{-3} = x^{2+(-3)} = x^{-1}$ or $\frac{1}{x}$

5) $\frac{x^8}{x^3} = x^{8-3} = x^5$

6) $\frac{x^2}{x^9} = x^{2-9} = x^{-7}$ or $\frac{1}{x^7}$

7) $\frac{x^6}{x^6} = 1$

8) $\frac{x^6}{x^{-6}} = x^{6-(-6)} = x^{6+6} = x^{12}$

9) $\frac{x^5}{x^{-3}} = x^{5-(-3)} = x^{5+3} = x^8$

10) $\frac{x^{-3}}{x^{-4}} = x^{-3-(-4)} = x^{-3+4} = x^1 = x$

11) $\frac{x^{-7}}{x} = \frac{x^{-7}}{x^1} = x^{-7-1} = x^{-8}$ or $\frac{1}{x^8}$

12) $(x^4)^5 = x^{4(5)} = x^{20}$

13) $(-x^6)^3 = (-1)^3 (x^6)^3 = -x^{18}$

14) $(-x^3)^8 = (-1)^8 (x^3)^8 = (1)x^{24} = x^{24}$

15) $(x^5)^{-2} = \frac{1}{(x^5)^2} = \frac{1}{x^{5(2)}} = \frac{1}{x^{10}}$ or x^{-10}

16) $(x^{1/3})^6 = x^{(1/3)6} = x^{6/3} = x^2$

17) $(4x^2)^3 = 4^3 (x^2)^3 = 64 x^{2(3)} = 64 x^6$

18) $(3x^4)^{-5} = \frac{1}{(3x^4)^5} = \frac{1}{3^5 (x^4)^5} = \frac{1}{243 x^{20}}$

19) $\sqrt{36x^2} = \sqrt{36}\sqrt{x^2} = \pm 6x$

20) $\sqrt{108 x^3} = \sqrt{36}\sqrt{x^2}\sqrt{3x} = \pm 6x\sqrt{3x}$
Alternate: $\pm 6x^{3/2}\sqrt{3}$

21) $\sqrt{\frac{25 x^{12}}{49}} = \frac{\sqrt{25}\sqrt{x^{12}}}{\sqrt{49}} = \pm\frac{5x^6}{7}$

22) $\sqrt{\frac{12 x^8}{27}} = \sqrt{\frac{4 x^8}{9}} = \frac{\sqrt{4}\sqrt{x^8}}{\sqrt{9}} = \pm\frac{2x^4}{3}$

Answer Key

Exercise Set 1.6

1) $\left(1+\frac{1}{10}\right)^{10} = (1+0.1)^{10} = (1.1)^{10} = 2.59374246$

2) $\left(1+\frac{1}{100}\right)^{100} = (1+0.01)^{100} = (1.01)^{100} = 2.704813829$

3) $\left(1+\frac{1}{1000}\right)^{1000} = (1+0.001)^{1000} = (1.001)^{1000} = 2.716923932$

4) $\left(1+\frac{1}{1,000,000}\right)^{1,000,000} = (1+0.000001)^{1,000,000} = (1.000001)^{1,000,000} = 2.718280469$

5) $\frac{1}{0!} + \frac{1}{1!} + \frac{1}{2!} + \frac{1}{3!} = 2.666666667$

6) $\frac{1}{0!} + \frac{1}{1!} + \frac{1}{2!} + \frac{1}{3!} + \frac{1}{4!} = 2.708333333$

7) $\frac{1}{0!} + \frac{1}{1!} + \frac{1}{2!} + \frac{1}{3!} + \frac{1}{4!} + \frac{1}{5!} = 2.716666667$

8) $\frac{1}{0!} + \frac{1}{1!} + \frac{1}{2!} + \frac{1}{3!} + \frac{1}{4!} + \frac{1}{5!} + \frac{1}{6!} = 2.718055556$

Question: In the limit that the number of terms of this series becomes infinite, the sum of the terms approaches Euler's number. We will explore this series in Chapter 9 (which is the chapter involving calculus).

Chapter 2 Logarithm Basics

Exercise Set 2.1

1) $\log_4 64$ asks, "Which exponent of 4 equals 64?"
The problem $y = \log_4 64$ is equivalent to $4^y = 64$.
The answer is $\log_4 64 = 3$ because $4^3 = 64$.
2) $\log_2 256$ asks, "Which exponent of 2 equals 256?"
The problem $y = \log_2 256$ is equivalent to $2^y = 256$.
The answer is $\log_2 256 = 8$ because $2^8 = 256$.
3) $\log_{10} 1,000,000$ asks, "Which exponent of 10 equals 1,000,000?"
The problem $y = \log_{10} 1,000,000$ is equivalent to $10^y = 1,000,000$.
The answer is $\log_{10} 1,000,000 = 6$ because $10^6 = 1,000,000$.
4) $\log_5 625$ asks, "Which exponent of 5 equals 625?"
The problem $y = \log_5 625$ is equivalent to $5^y = 625$.
The answer is $\log_5 625 = 4$ because $5^4 = 625$.

Exercise Set 2.2

1) $\log_{10} 0.00001$ asks, "Which exponent of 10 equals 0.00001?"
The problem $y = \log_{10} 0.00001$ is equivalent to $10^y = 0.00001$.
The answer is $\log_{10} 0.00001 = -5$ because $10^{-5} = \left(\frac{1}{10}\right)^5 = \frac{1}{10^5} = 0.00001$.
2) $\log_{25} 0.04$ asks, "Which exponent of 25 equals 0.04?"
The problem $y = \log_{25} 0.04$ is equivalent to $25^y = 0.04$.
The answer is $\log_{25} 0.04 = -1$ because $25^{-1} = \frac{1}{25} = \frac{1 \times 4}{25 \times 4} = \frac{4}{100} = 0.04$.

Exercise Set 2.3

1) $\log_{10} 10,000 = 4$ because $10^4 = 10,000$ (there are 4 zeros)
2) $\log_{10} 1,000,000 = 6$ because $10^6 = 1,000,000$ (there are 6 zeros)
3) $\log_{10} 100 = 2$ because $10^2 = 100$ (there are 2 zeros)

Answer Key

4) $\log_{10} 0.000001 = -6$ because $10^{-6} = 0.000001$ (there are 5 zeros between the decimal point and the 1, and $5 + 1 = 6$; it is negative because $0.000001 < 1$)

5) $\log_{10} 0.1 = -1$ because $10^{-1} = 0.1$ (there are no zeros between the decimal point and the 1, and $0 + 1 = 1$; it is negative because $0.1 < 1$)

6) $\log_{10} 100,000,000 = 8$ because $10^8 = 100,000,000$ (there are 8 zeros)

7) $\log_{10} 1 = 0$ because $10^0 = 1$ (there are no zeros)

8) $\log_{10} 0.01 = -2$ because $10^{-2} = 0.01$ (there is 1 zero between the decimal point and the 1, and $1 + 1 = 2$; it is negative because $0.01 < 1$)

9) $\log_{10} 1,000,000,000 = 9$ because $10^9 = 1,000,000,000$ (there are 9 zeros)

10) $\log_{10} 0.0000001 = -7$ because $10^{-7} = 0.0000001$ (there are 6 zeros between the decimal point and the 1, and $6 + 1 = 7$; it is negative because $0.0000001 < 1$)

11) $\log_{10} 1,000,000,000,000 = 12$ because $10^{12} = 1,000,000,000,000$ (there are 12 zeros)

12) $\log_{10} 10,000,000,000 = 10$ because $10^{10} = 10,000,000,000$ (there are 10 zeros)

13) $\log_{10} 0.0000000001 = -10$ because $10^{-10} = 0.0000000001$ (there are 9 zeros between the decimal point and the 1, and $9 + 1 = 10$; it is negative because the argument is less than 1)

14) $\log_{10} 0.000000000001 = -12$ because $10^{-12} = 0.000000000001$ (there are 11 zeros between the decimal point and the 1, and $11 + 1 = 12$; it is negative because the argument is less than 1)

15) $\log_{10}(10^{38}) = 38$ because $y = \log_{10} 10^{38}$ is equivalent to $10^{38} = 10^y$ (because in general $y = \log_{10} x$ is equivalent to $x = 10^y$; so if $x = 10^{38}$, we get $y = 38$)

16) $\log_{10}(10^{-75}) = -75$ because $y = \log_{10} 10^{-75}$ is equivalent to $10^{-75} = 10^y$ (because in general $y = \log_{10} x$ is equivalent to $x = 10^y$; so if $x = 10^{-75}$, we get $y = -75$)

Exercise Set 2.4

1) $\log_3 81 = 4$ because $3^4 = 81$

2) $\log_4 1024 = 5$ because $4^5 = 1024$

3) $\log_5 125 = 3$ because $5^3 = 125$

4) $\log_{100} 1,000,000,000,000 = 6$ because $100^6 = 1,000,000,000,000$ (base 100, **not** 10)

5) $\log_6 \left(\frac{1}{36}\right) = -2$ because $6^{-2} = \frac{1}{6^2} = \frac{1}{36}$

Logarithms and Exponentials Essential Skills Practice Workbook with Answers

6) $\log_2 0.0625 = -4$ because $2^{-4} = \frac{1}{2^4} = \frac{1}{16} = 0.0625$

7) $\log_7 1 = 0$ because $7^0 = 1$

8) $\log_{100} 0.000001 = -3$ because $100^{-3} = \frac{1}{100^3} = \frac{1}{1000000} = 0.000001$ (base 100, **not** 10)

9) $\log_8 0.125 = -1$ because $8^{-1} = \frac{1}{8} = 0.125$

10) $\log_5 0.008 = -3$ because $5^{-3} = \frac{1}{5^3} = \frac{1}{125} = 0.008$

11) $\log_{20} 160,000 = 4$ because $20^4 = 160,000$

12) $\log_2 2048 = 11$ because $2^{11} = 2048$

13) $\log_3 729 = 6$ because $3^6 = 729$

14) $\log_{25} 625 = 2$ because $25^2 = 625$

15) $\log_{20} 0.000125 = -3$ because $20^{-3} = \frac{1}{20^3} = \frac{1}{8000} = 0.000125$

16) $\log_{25} 0.0016 = -2$ because $25^{-2} = \frac{1}{25^2} = \frac{1}{625} = 0.0016$

17) $\log_4 0.015625 = -3$ because $4^{-3} = \frac{1}{4^3} = \frac{1}{64} = 0.015625$

18) $\log_3 \frac{1}{81} = -4$ because $3^{-4} = \frac{1}{3^4} = \frac{1}{81}$

19) $\log_4 32 = \frac{5}{2} = 2.5$ because $4^{2.5} = 4^{5/2} = (\sqrt{4})^5 = 2^5 = 32$

20) $\log_{27} 81 = \frac{4}{3}$ because $27^{4/3} = (\sqrt[3]{27})^4 = 3^4 = 81$

21) $\log_{100} 10 = \frac{1}{2} = 0.5$ because $100^{0.5} = 100^{1/2} = (\sqrt{100})^1 = 10$ (base 100, **not** 10)

22) $\log_{16} 512 = \frac{9}{4} = 2.25$ because $16^{2.25} = 16^{9/4} = (\sqrt[4]{16})^9 = 2^9 = 512$

23) $\log_9 243 = \frac{5}{2} = 2.5$ because $9^{2.5} = 9^{5/2} = (\sqrt{9})^5 = 3^5 = 243$

24) $\log_8 4 = \frac{2}{3}$ because $8^{2/3} = (\sqrt[3]{8})^2 = 2^2 = 4$

25) $\log_{25} 0.00032 = -\frac{5}{2} = -2.5$ because $25^{-2.5} = 25^{-5/2} = \frac{1}{25^{5/2}} = \frac{1}{(\sqrt{25})^5} = \frac{1}{5^5} = \frac{1}{3125} = 0.00032$

26) $\log_{1000} 0.01 = -\frac{2}{3}$ because $1000^{-2/3} = \frac{1}{1000^{2/3}} = \frac{1}{(\sqrt[3]{1000})^2} = \frac{1}{10^2} = \frac{1}{100} = 0.01$

27) $\log_{27}\left(\frac{1}{9}\right) = -\frac{2}{3}$ because $27^{-2/3} = \frac{1}{27^{2/3}} = \frac{1}{(\sqrt[3]{27})^2} = \frac{1}{3^2} = \frac{1}{9}$

28) $\log_{64} 0.25 = -\frac{1}{3}$ because $64^{-1/3} = \frac{1}{64^{1/3}} = \frac{1}{(\sqrt[3]{64})^1} = \frac{1}{4^1} = \frac{1}{4} = 0.25$

29) $\log_{81}\left(\frac{1}{243}\right) = -\frac{5}{4} = -1.25$ because $81^{-1.25} = 81^{-5/4} = \frac{1}{81^{5/4}} = \frac{1}{\left(\sqrt[4]{81}\right)^5} = \frac{1}{3^5} = \frac{1}{243}$

30) $\log_{32}\left(\frac{1}{16}\right) = -\frac{4}{5} = -0.8$ because $32^{-0.8} = 32^{-4/5} = \frac{1}{32^{4/5}} = \frac{1}{\left(\sqrt[5]{32}\right)^4} = \frac{1}{2^4} = \frac{1}{16}$

Exercise Set 2.5

1) $\ln 5 \approx 1.61$

2) $\ln 2 \approx 0.693$

3) $\ln 10 \approx 2.30$

4) $\ln 0.5 \approx -0.693$

5) $\ln 1 = 0$ (it's exact)

6) $\ln 0.999 \approx -0.00100$

7) $\ln 25 \approx 3.22$

8) $\ln 100 \approx 4.61$

9) $\ln 0.367879 \approx -1.00$

10) $\ln 0.007 \approx -4.96$

11) $\ln 489 \approx 6.19$

12) $\ln 23{,}492 \approx 10.1$

13) $\ln 0.000001 \approx -13.8$

14) $\ln 10^{15} \approx 34.5$

15) $\ln\left(\frac{2}{3}\right) \approx -0.405$

16) $\ln \sqrt{2} \approx 0.347$

17) $\ln(49^3) \approx 11.7$

18) $\ln(8^{-3}) \approx -6.24$

19) $\ln(e^5) = 5.00$ (it's exact)

20) $\ln\left(\frac{1}{e^8}\right) = -8.00$ (it's exact)

Chapter 3 Logarithm Rules

Exercise Set 3.1

1) $\log_5(5^8) = 8$
2) $\log_{10}(10^{0.4}) = 0.4$
3) $\ln(e^9) = 9$
4) $\log_3(3^{-4/9}) = -\frac{4}{9}$
5) $6^{\log_6 3} = 3$
6) $42^{\log_{42}(2/3)} = \frac{2}{3}$
7) $e^{\ln 2} = 2$
8) $10^{\log_{10} 4.17} = 4.17$
9) $\log_{10}(10^{123}) = 123$
10) $\ln(e^0) = 0$
11) $\ln(e^{\sqrt{2}}) = \sqrt{2}$
12) $\log_7(7^{1/7}) = \frac{1}{7}$
13) $17^{\log_{17} 0.001} = 0.001$
14) $e^{\ln 0.8} = 0.8$
15) $e^{\ln 1} = 1$
16) $9^{\log_9 e} = e$
17) $\log_2[2^{(3^4)}] = 3^4 = 81$
18) $\ln(e^{-1/\pi}) = -\frac{1}{\pi}$
19) $t^{\log_t \sqrt{3}} = \sqrt{3}$
20) $\log_x(x^{-5.2}) = -5.2$

Exercise Set 3.2

1) $\log_2(0.2) + \log_2 320 = \log_2(0.2 \times 320) = \log_2 64 = 6$ since $2^6 = 64$
2) $\log_{10} 40 + \log_{10} 25 = \log_{10}(40 \times 25) = \log_{10} 1000 = 3$ since $10^3 = 1000$
3) $\log_6 4 + \log_6 9 = \log_6(4 \times 9) = \log_6 36 = 2$ since $6^2 = 36$
4) $\ln(0.25e) + \ln(4e) = \ln(0.25e \times 4e) = \ln(e^2) = 2$ (using the cancellation equation)
5) $\log_4 2 + \log_4 \sqrt{2} + \log_4 32 + \log_4 \sqrt{8} = \log_4(2 \times \sqrt{2} \times 32 \times \sqrt{8}) = \log_4(64 \times \sqrt{16})$
$= \log_4(64 \times 4) = \log_4 256 = 4$ since $4^4 = 256$
Note: Recall from Sec. 1.5 that $\sqrt{ax} = \sqrt{a}\sqrt{x}$, such that $\sqrt{2}\sqrt{8} = \sqrt{2 \times 8} = \sqrt{16} = 4$.
6) $\log_{10}(3x) + \log_{10}(4x) = \log_{10}[(3x)(4x)] = \log_{10}(12x^2)$
7) $\log_2(x-3) + \log_2(x+3) = \log_2[(x-3)(x+3)] = \log_2(x^2-9)$
8) $\ln x + \ln(2x) + \ln(3x) = \ln[x(2x)(3x)] = \ln(6x^3)$
9) $\log_{10}|\tan x| + \log_{10}|\cos x| = \log_{10}|\tan x \cos x| = \log_{10}\left|\frac{\sin x}{\cos x} \cos x\right| = \log_{10}|\sin x|$
10) $\ln \sqrt{7x} + \ln \sqrt{7x} = \ln(\sqrt{7x}\sqrt{7x}) = \ln(\sqrt{7x \cdot 7x}) = \ln(\sqrt{49x^2}) = \ln(7x)$

Answer Key

Exercise Set 3.3

1) $\log_{10} 2500 - \log_{10}(2.5) = \log_{10}\left(\frac{2500}{2.5}\right) = \log_{10}(1000) = 3$ since $10^3 = 1000$

2) $\log_3 486 - \log_3 6 = \log_3\left(\frac{486}{6}\right) = \log_3(81) = 4$ since $3^4 = 81$

3) $\ln(7e) - \ln 7 = \ln\left(\frac{7e}{7}\right) = \ln e = 1$

4) $\log_3 \sqrt{45} - \log_3 \sqrt{5} = \log_3\left(\frac{\sqrt{45}}{\sqrt{5}}\right) = \log_3 \sqrt{\frac{45}{5}} = \log_3 \sqrt{9} = \log_3 3 = 1$ since $3^1 = 3$

5) $\log_6 1 - \log_6 4 - \log_6 9 = \log_6\left(\frac{1}{4 \cdot 9}\right) = \log_6\left(\frac{1}{36}\right) = -2$ since $6^{-2} = \frac{1}{6^2} = \frac{1}{36}$

6) $\log_5(24x^8) - \log_5(6x^3) = \log_5\left(\frac{24x^8}{6x^3}\right) = \log_5(4x^5)$

7) $\ln(72x) - \ln(4x) - \ln(3x) = \ln\left(\frac{72x}{4x \cdot 3x}\right) = \ln\left(\frac{72x}{12x^2}\right) = \ln\left(\frac{6}{x}\right)$

8) $\log_{10}(x-5) - \log_{10}(x+2) - \log_{10}(x-2) = \log_{10}\left[\frac{x-5}{(x+2)(x-2)}\right] = \log_{10}\left(\frac{x-5}{x^2-4}\right)$

9) $\ln|\cos x| - \ln|\sin x| = \ln\left|\frac{\cos x}{\sin x}\right| = \ln|\cot x|$

10) $\log_2 \sqrt{48x} - \log_2 \sqrt{3x} = \log_2\left(\frac{\sqrt{48x}}{\sqrt{3x}}\right) = \log_2 \sqrt{\frac{48x}{3x}} = \log_2 \sqrt{16} = \log_2 4 = 2$ since $2^2 = 4$

Exercise Set 3.4

1) $\log_2(8^{7.5}) = 7.5 \log_2 8 = 7.5(3) = 22.5$ since $2^3 = 8$

2) $\log_{10}(0.1^{3/2}) = \frac{3}{2}\log_{10}(0.1) = \frac{3}{2}(-1) = -\frac{3}{2} = -1.5$ since $10^{-1} = 0.1$

3) $\log_6(36^{0.3}) = 0.3 \log_6 36 = 0.3(2) = 0.6$ since $6^2 = 36$

4) $\ln(e^\pi) = \pi \ln e = \pi$ since $\ln e = 1$

5) $\log_3 \sqrt{243} = \log_3[(243)^{1/2}] = \frac{1}{2}\log_3 243 = \frac{1}{2}(5) = \frac{5}{2} = 2.5$ since $3^5 = 243$

6) $\log_{10}(x^{5/3}) = \frac{5}{3}\log_{10} x$

7) $5\ln(x^{4.2}) = 5(4.2)\ln x = 21 \ln x$

8) $\log_2(6^5 x^5) = \log_2[(6x)^5] = 5\log_2(6x)$

9) $\log_{10}(\sin^2 x) = 2\log_{10}|\sin x|$ Note: the absolute values reflect that logarithms only give real values when the argument is positive.

10) $\ln \sqrt{5x} = \ln[(5x)^{1/2}] = \frac{1}{2}\ln(5x)$

Logarithms and Exponentials Essential Skills Practice Workbook with Answers

Exercise Set 3.5

1) $\log_{10}(10^{-1}) = -\log_{10} 10 = -1$ since $10^1 = 10$

2) $\log_9\left(\frac{1}{81}\right) = \log_9(81^{-1}) = -\log_9 81 = -2$ since $9^2 = 81$

3) $\log_6(6^{-2}) = -\log_6(6^2) = -\log_6 36 = -2$ since $6^2 = 36$

4) $\ln(e^{-3}) = -\ln(e^3) = -3\ln e = -3(1) = -3$ since $\ln e = 1$

5) $\log_4\left(\frac{1}{64}\right) = \log_4(64^{-1}) = -\log_4 64 = -3$ since $4^3 = 64$

6) $\log_2(3^{-1}x^{-1}) = \log_2[(3x)^{-1}] = -\log_2(3x)$

7) $-6\ln\left(\frac{1}{2x}\right) = -6\ln[(2x)^{-1}] = (-1)(-6)\ln(2x) = 6\ln(2x)$

8) $\log_{10}\left(\frac{1}{3xy}\right) = \log_{10}[(3xy)^{-1}] = -\log_{10}(3xy)$

9) $\log_{10}\left|\frac{1}{\cos x}\right| = \log_{10}|\cos^{-1} x| = -\log_{10}|\cos x|$

10) $\ln\left(\frac{1}{4x} + \frac{3}{4x}\right) = \ln\left(\frac{1+3}{4x}\right) = \ln\left(\frac{4}{4x}\right) = \ln\left(\frac{1}{x}\right) = \ln(x^{-1}) = -\ln x$

Note: Add fractions by making a common denominator. Since the denominator is already the same, we may add the numerators.

Exercise Set 3.6

1) $\log_6 54 + \log_6 12 - \log_6 18 = \log_6\left[\frac{(54)(12)}{18}\right] = \log_6 36 = 2$ since $6^2 = 36$

2) $2\log_3 2 - 2\log_3 6 = \log_3(2^2) - \log_3(6^2) = \log_3 4 - \log_3 36 = \log_3\left(\frac{4}{36}\right) = \log_3\left(\frac{1}{9}\right) = -2$
since $3^{-2} = \frac{1}{3^2} = \frac{1}{9}$

3) $8e^{-3\ln 4} = 8e^{\ln(4^{-3})} = 8(4^{-3}) = \frac{8}{4^3} = \frac{8}{64} = \frac{1}{8}$ or 0.125

4) $12\log_4(2^3) = \log_4[(2^3)^{12}] = \log_4(2^{36}) = \log_4[(2^2)^{18}] = \log_4(4^{18}) = 18$

5) $5\ln\left[\left(\frac{1}{e^5}\right)^8\right] = 5\ln[(e^{-5})^8] = 5\ln(e^{-40}) = 5(-40) = -200$

6) $\log_{10}(100^{6\log_3 9}) = \log_{10}[100^{6\log_3(3^2)}] = \log_{10}[100^{\log_3(3^{12})}] = \log_{10}(100^{12})$
$= \log_{10}[(10^2)^{12}] = \log_{10}(10^{24}) = 24$

Answer Key

7) $36 \log_{100} \sqrt{\frac{1}{10}} = 36 \log_{100} \sqrt{10^{-1}} = 36 \log_{100}[(10^{-1})^{1/2}] = 36 \log_{100}(10^{-1/2})$
$= \log_{100}(10^{-18}) = \log_{100}[(10^2)^{-9}] = \log_{100}(100^{-9}) = -9$ (base 100, **not** 10)

8) $\log_6 \left(\frac{2}{3}\right) + \log_6 \left(\frac{1}{24}\right) = \log_6 \left(\frac{2}{3} \times \frac{1}{24}\right) = \log_6 \left(\frac{2}{72}\right) = \log_6 \left(\frac{1}{36}\right) = -2$ since $6^{-2} = \frac{1}{6^2} = \frac{1}{36}$

9) $6 \ln[\ln(e^{e^{-5}})] = 6 \ln(e^{-5}) = 6(-5) = -30$

If it helps, first define $y = e^{-5}$ to get $6 \ln[\ln(e^y)] = 6 \ln y$.

10) $3 \log_2 \sqrt{8} + 5 \log_2 \sqrt{32} = 3 \log_2(8^{1/2}) + 5 \log_2(32^{1/2}) = \log_2(8^{3/2}) + \log_2(32^{5/2})$
$= \log_2[(2^3)^{3/2}] + \log_2[(2^5)^{5/2}] = \log_2(2^{9/2}) + \log_2(2^{25/2}) = \log_2[(2^{9/2})(2^{25/2})]$
$= \log_2(2^{9/2+25/2}) = \log_2(2^{34/2}) = \log_2(2^{17}) = 17$

Note: In Chapter 4, you'll learn how you can check these answers with a calculator. In case you may have any doubts about the answers, you'll be able to use a calculator to check them.

11) $4 \log_7 x^2 - 3 \log_7 x = 4(2) \log_7 x - 3 \log_7 x = 8 \log_7 x - 3 \log_7 x$
$= (8-3) \log_7 x = 5 \log_7 x$

12) $\log_5(x+5) + \log_5(x-5) - \log_5(x^2-25) = \log_5\left[\frac{(x+5)(x-5)}{x^2-25}\right] = \log_5\left(\frac{x^2-25}{x^2-25}\right)$
$= \log_5 1 = 0$

13) $5 \ln\left(\frac{4}{x}\right) + 7 \ln\left(\frac{x}{2}\right) + \ln x = \ln\left[\left(\frac{4}{x}\right)^5\right] + \ln\left[\left(\frac{x}{2}\right)^7\right] + \ln x = \ln\left(\frac{4^5}{x^5}\right) + \ln\left(\frac{x^7}{2^7}\right) + \ln x$
$= \ln\left(\frac{4^5 x^7 x}{2^7 x^5}\right) = \ln\left(\frac{1024 x^8}{128 x^5}\right) = \ln(8x^3) = \ln(2^3 x^3) = \ln[(2x)^3] = 3 \ln(2x)$

14) $\log_{10}\left(\frac{x}{y}\right) - \log_{10}(4x) = \log_{10}\left(\frac{x}{y} \div 4x\right) = \log_{10}\left(\frac{x}{y} \cdot \frac{1}{4x}\right) = \log_{10}\left(\frac{x}{4xy}\right) = \log_{10}\left(\frac{1}{4y}\right)$
$= \log_{10}[(4y)^{-1}] = -\log_{10}(4y)$

15) $6 \log_3 x^3 - 3 \log_3 x^4 + 4 \log_3 x^2 = \log_3[(x^3)^6] - \log_3[(x^4)^3] + \log_3[(x^2)^4]$
$= \log_3(x^{18}) - \log_3(x^{12}) + \log_3(x^8) = \log_3\left(\frac{x^{18} x^8}{x^{12}}\right) = \log_3\left(\frac{x^{26}}{x^{12}}\right) = \log_3(x^{14}) = 14 \log_3 x$

16) $4x^6 e^{-5 \ln x} = 4x^6 e^{\ln(x^{-5})} = 4x^6 x^{-5} = 4x$

17) $\log_2(x^{1/2}) - \frac{1}{3} \log_2 x = \log_2(x^{1/2}) - \log_2(x^{1/3}) = \log_2\left(\frac{x^{1/2}}{x^{1/3}}\right) = \log_2(x^{1/2 - 1/3})$
$= \log_2(x^{1/6}) = \frac{1}{6} \log_2 x$

18) $\ln(\sin^2 x) - \ln(\cos^2 x) = \ln\left(\frac{\sin^2 x}{\cos^2 x}\right) = \ln(\tan^2 x) = 2 \ln|\tan x|$ Note: the absolute values reflect that logarithms only give real values when the argument is positive.

Logarithms and Exponentials Essential Skills Practice Workbook with Answers

19) $3\log_{10}\left(\frac{x^3}{y^4}\right) - 6\log_{10}\left(\frac{x}{y^2}\right) = \log_{10}\left[\left(\frac{x^3}{y^4}\right)^3\right] - \log_{10}\left[\left(\frac{x}{y^2}\right)^6\right] = \log_{10}\left(\frac{x^9}{y^{12}}\right) - \log_{10}\left(\frac{x^6}{y^{12}}\right)$

$= \log_{10}\left(\frac{x^9}{y^{12}} \div \frac{x^6}{y^{12}}\right) = \log_{10}\left(\frac{x^9}{y^{12}} \cdot \frac{y^{12}}{x^6}\right) = \log_{10}(x^3) = 3\log_{10}x$

20) $\ln\sqrt{x^5} + \ln\sqrt{x^3} - 6\ln(x^2) = \ln(\sqrt{x^5}\sqrt{x^3}) - 6\ln(x^2) = \ln(\sqrt{x^5 x^3}) - 6\ln(x^2)$

$= \ln(\sqrt{x^8}) - 6\ln(x^2) = \ln(x^4) - 6\ln(x^2) = \ln(x^4) - \ln[(x^2)^6] = \ln(x^4) - \ln(x^{12})$

$= \ln\left(\frac{x^4}{x^{12}}\right) = \ln(x^{-8}) = -8\ln x$

Exercise Set 3.7

1) Divide both sides by 6 to get $\log_5 x = 3$

Exponentiate both sides (with base 5) to get $5^{\log_5 x} = 5^3$

Apply the cancellation equation $b^{\log_b x} = x$ to get $x = 5^3 = \boxed{125}$

Check: $6\log_5 x = 6\log_5 125 = 6(3) = 18$

2) Add 1 to both sides to get $2\log_9 x = 4$

Divide both sides by 2 to get $\log_9 x = 2$

Exponentiate both sides (with base 9) to get $9^{\log_9 x} = 9^2$

Apply the cancellation equation $b^{\log_b x} = x$ to get $x = 9^2 = \boxed{81}$

Check: $2\log_9 x - 1 = 2\log_9 81 - 1 = 2(2) - 1 = 4 - 1 = 3$

3) Subtract 4 from both sides to get $-\log_2 x = 5$

Multiply both sides by -1 to get $\log_2 x = -5$

Exponentiate both sides (with base 2) to get $2^{\log_2 x} = 2^{-5}$

Apply the cancellation equation $b^{\log_b x} = x$ to get $x = 2^{-5} = \frac{1}{2^5} = \boxed{\frac{1}{32}}$

Check: $4 - \log_2 x = 4 - \log_2\left(\frac{1}{32}\right) = 4 - (-5) = 4 + 5 = 9$

4) Exponentiate both sides (with base 3) to get $3^{\log_3(x/2)} = 3^4$

Apply the cancellation equation $b^{\log_b(x/2)} = \frac{x}{2}$ to get $\frac{x}{2} = 3^4 = 81$

Multiply both sides by 2 to get $x = 2(81) = \boxed{162}$

Check: $\log_3\left(\frac{x}{2}\right) = \log_3\left(\frac{162}{2}\right) = \log_3(81) = 4$

Answer Key

5) Subtract 2 from both sides to get $\log_{10}(x^2) = 6$

Exponentiate both sides (with base 10) to get $10^{\log_{10}(x^2)} = 10^6$

Apply the cancellation equation $b^{\log_b(x^2)} = x^2$ to get $x^2 = 10^6 = 1{,}000{,}000$

Square root both sides to get $x = \sqrt{1{,}000{,}000} = \boxed{\pm 1000}$

Check: $\log_{10}(x^2) + 2 = \log_{10}[(\pm 1000)^2] + 2 = \log_{10} 1{,}000{,}000 + 2 = 6 + 2 = 8$

6) Use the difference of logarithms formula to get $\log_6\left(\frac{x^5}{x^2}\right) = 6$

Simplify the left side to get $\log_6(x^3) = 6$

Exponentiate both sides (with base 6) to get $6^{\log_6 x} = 6^6$

Apply the cancellation equation $b^{\log_b(x^3)} = x^3$ to get $x^3 = 6^6$

Cube root both sides to get $x = (6^6)^{1/3} = 6^2 = 36$

Alternative solution: $\log_6(x^5) - \log_6(x^2) = 5\log_6 x - 2\log_6 x = 3\log_6 x = 6$

Divide both sides by 3 to get $\log_6 x = 2$

Exponentiate both sides (with base 6) to get $6^{\log_6 x} = 2$

Apply the cancellation equation $b^{\log_b x} = x$ to get $x = 6^2 = \boxed{36}$

Check: $\log_6(x^5) - \log_6(x^2) = 5\log_6(36) - 2\log_6(36) = 5(2) - 2(2) = 10 - 4 = 6$

7) Use the sum of logarithms formula to get $\log_4\left(\sqrt{x}\sqrt{x}\right) = 5$

Simplify the left side to get $\log_4\left(\sqrt{x^2}\right) = \log_4 x = 5$

Exponentiate both sides (with base 4) to get $4^{\log_4 x} = 4^5$

Apply the cancellation equation $b^{\log_b x} = x$ to get $x = 4^5 = \boxed{1024}$

Check: $\log_4 \sqrt{1024} + \log_4 \sqrt{1024} = \log_4 32 + \log_4 32 = \frac{5}{2} + \frac{5}{2} = \frac{10}{2} = 5$

Note: $4^{5/2} = \left(\sqrt{4}\right)^5 = 2^5 = 32$

8) Exponentiate both sides (with base 9) to get $9^{\log_9(6-3x)} = 9^2$

Apply the cancellation equation $b^{\log_b(6-3x)} = 6 - 3x$ to get $6 - 3x = 9^2 = 81$

Subtract 6 from both sides to get $-3x = 75$

Divide by -3 on both sides to get $x = \boxed{-25}$

Check: $\log_9(6 - 3x) = \log_9[6 - 3(-25)] = \log_9(6 + 75) = \log_9 81 = 2$

Logarithms and Exponentials Essential Skills Practice Workbook with Answers

9) Use the sum of logarithms formula to get $\log_3[(x+3)(x-3)] = 3$

Simplify the left side to get $\log_3(x^2 - 9) = 3$

Exponentiate both sides (with base 3) to get $3^{\log_3(x^2-9)} = 3^3$

Apply the cancellation equation $b^{\log_b(x^2-9)} = x^2 - 9$ to get $x^2 - 9 = 3^3 = 27$

Add 9 to both sides to get $x^2 = 36$

Square root both sides to get $x = \sqrt{36} = \boxed{6}$ (only the positive root works; if you try the negative root, you will get a negative argument for one logarithm, which is a problem)

Check: $\log_3(x+3) + \log_3(x-3) = \log_3(6+3) + \log_3(6-3) = \log_3 9 + \log_3 3 = 2 + 1 = 3$

10) Use the difference of logarithms formula to get $\log_2\left(\frac{5x-6}{2x+3}\right) = 0$

Exponentiate both sides (with base 2) to get $2^{\log_6\left(\frac{5x-6}{2x+3}\right)} = 2^0$

Apply the cancellation equation $b^{\log_b\left(\frac{5x-6}{2x+3}\right)} = \frac{5x-6}{2x+3}$ to get $\frac{5x-6}{2x+3} = 2^0 = 1$

Multiply by $2x - 3$ on both sides to get $5x - 6 = (1)(2x+3)$

Simplify the right side to get $5x - 6 = 2x + 3$

Add 6 to both sides to get $5x = 2x + 9$

Subtract $2x$ from both sides to get $3x = 9$

Divide by 3 on both sides to get $x = \boxed{3}$

Check: $\log_2(5x - 6) - \log_2(2x + 3) = \log_2[5(3) - 6] - \log_2[2(3) + 3]$
$= \log_2(15 - 6) - \log_2(6 + 3) = \log_2 9 - \log_2 9 = \log_2\left(\frac{9}{9}\right) = \log_2 1 = 0$

11) Divide by 4 on both sides to get $\ln x = 3$

Exponentiate both sides (with base e) to get $e^{\ln x} = e^3$

Apply the cancellation equation $e^{\ln x} = x$ to get $x = e^3$

Use a calculator's EXP or e^x button to determine that $x = \boxed{e^3} \approx \boxed{20.0855}$

Check: $4 \ln x = 4 \ln(e^3) = 4(3) = 12$

12) Use the difference of logarithms formula to get $\ln\left(\frac{6x^4}{2x^2}\right) = 3$

Simplify the left side to get $\ln(3x^2) = 3$

Exponentiate both sides (with base e) to get $e^{\ln(3x^2)} = e^3$

Apply the cancellation equation $e^{\ln(3x^2)} = 3x^2$ to get $3x^2 = e^3$

Answer Key

Divide both sides by 3 to get $x^2 = \frac{e^3}{3}$

Square root both sides to get $x = \sqrt{\frac{e^3}{3}}$

Use a calculator to determine that $x = \boxed{\sqrt{\frac{e^3}{3}}} \approx \boxed{2.5875}$

Check: $\ln(6x^4) - \ln(2x^2) = \ln[6(2.5875)^4] - \ln[2(2.5875)^2] \approx 3$

13) Use the sum of logarithms formula to get $\ln[x(3x + 2)] = 1$

Distribute on the left side to get $\ln(3x^2 + 2x) = 1$

Exponentiate both sides (with base e) to get $e^{\ln(3x^2+2x)} = e^1$

Apply the cancellation equation $e^{\ln(3x^2+2x)} = 3x^2 + 2x$ to get $3x^2 + 2x = e$

Subtract e from both sides to get $3x^2 + 2x - e = 0$

This is a quadratic equation. Compare the general equation $ax^2 + bx + c = 0$ to $3x^2 + 2x - e = 0$ to see that $a = 3$, $b = 2$, and $c = -e$. Plug these values into the quadratic formula. Use a calculator.

$$x = \frac{-b \pm \sqrt{b^2 - 4ac}}{2a} = \frac{-2 \pm \sqrt{2^2 - 4(3)(-e)}}{2(3)} = \frac{-2 \pm \sqrt{4 + 12e}}{6}$$

$$x \approx \frac{-2 \pm \sqrt{4 + 12(2.71828)}}{6} = \frac{-2 \pm 6.05139}{6}$$

The positive answer is $x = \boxed{0.675232}$ (the negative answer would pose a problem for the first logarithm, giving it a negative argument).

Check: $\ln x + \ln(3x + 2) \approx \ln(0.675232) + \ln[3(0.675232) + 2] \approx 1$

14) It may help to define $u = \ln x$, so that the problem becomes $\ln y = 1$.

Exponentiate both sides (with base e) to get $e^{\ln u} = e^1$

Apply the cancellation equation $e^{\ln u} = u$ to get $u = e^1 = e$

Note that this is u, not x. We're not finished. Plug in $u = \ln x$ to get $\ln x = e$.

Exponentiate both sides (with base e) to get $e^{\ln x} = e^e$

Apply the cancellation equation $e^{\ln x} = x$ to get $x = \boxed{e^e} \approx \boxed{15.154}$

Alternate solution: Exponentiate both sides (with base e) to get $e^{\ln(\ln x)} = e^1$

Apply the cancellation equation $e^{\ln(\ln x)} = \ln x$ to get $\ln x = e^1 = e$

Exponentiate again to get $e^{\ln x} = e^e$, which simplifies to $x = e^e$

Check: $\ln(\ln x) = \ln[\ln(e^e)] = \ln e = 1$

Logarithms and Exponentials Essential Skills Practice Workbook with Answers

15) Exponentiate both sides (with base e) to get $e^{\ln|\sin x|} = e^{-1}$

Apply the cancellation equation $e^{\ln|\sin x|} = |\sin x|$ to get $|\sin x| = e^{-1} = \dfrac{1}{e}$

Take the inverse sine of both sides to get $x = \boxed{\sin^{-1}\left(\dfrac{1}{e}\right)}$

Use a calculator to determine that $x \approx \boxed{0.3767}$ radians (about $\boxed{21.58}$ degrees).

Check: $\ln|\sin x| \approx \ln|\sin(0.3767 \text{ rad})| \approx \ln|\sin(21.58°)| \approx -1$

Note: Don't confuse $\dfrac{1}{e} \approx 0.3679$ with $\sin^{-1}\left(\dfrac{1}{e}\right) \approx 0.3767$. These numbers are close, but different. (It's a consequence of the property that the sine of an angle approximately equals the angle in radians for small angles.)

Answer Key

Chapter 4 Change of Base

Exercise Set 4.1

1) Plug $b = 9$, $x = 5$, and $a = 10$ into the change of base formula.

$$\log_9 5 = \frac{\log_a x}{\log_a b} = \frac{\log_{10} 5}{\log_{10} 9}$$

2) Plug $b = 2$ and $x = \frac{1}{3}$ into the change of base formula, using natural logarithms for the new base.

$$\log_2\left(\frac{1}{3}\right) = \frac{\ln x}{\ln b} = \frac{\ln\left(\frac{1}{3}\right)}{\ln 2}$$

3) Plug $b = 5$, $x = \pi$, and $a = 2$ into the change of base formula.

$$\log_5 \pi = \frac{\log_a x}{\log_a b} = \frac{\log_2 \pi}{\log_2 5}$$

Exercise Set 4.2

1) Plug $b = 3$, $x = 81$, and $a = 9$ into the change of base formula.

$$\log_3 81 = \frac{\log_a x}{\log_a b} = \frac{\log_9 81}{\log_9 3}$$

- The left side equals $\log_3 81 = \boxed{4}$ because $3^4 = 81$.
- The right side equals $\frac{\log_9 81}{\log_9 3} = \frac{2}{1/2} = 4$ because $9^2 = 81$ and $9^{1/2} = \sqrt{9} = 3$. To divide by a fraction, multiply by its reciprocal: $\frac{2}{1/2} = 2 \div \frac{1}{2} = 2 \times \frac{2}{1} = \frac{4}{1} = 4$.
- Both sides equal 4.

2) Plug $b = 8$, $x = 512$, and $a = 2$ into the change of base formula.

$$\log_8 512 = \frac{\log_a x}{\log_a b} = \frac{\log_2 512}{\log_2 8}$$

- The left side equals $\log_8 512 = \boxed{3}$ because $8^3 = 512$.
- The right side equals $\frac{\log_2 512}{\log_2 8} = \frac{9}{3} = 3$ because $2^9 = 512$ and $2^3 = 8$.
- Both sides equal 3.

3) Plug $b = 16$, $x = 256$, and $a = 4$ into the change of base formula.

$$\log_{16} 256 = \frac{\log_a x}{\log_a b} = \frac{\log_4 256}{\log_4 16}$$

- The left side equals $\log_{16} 256 = \boxed{2}$ because $16^2 = 256$.
- The right side equals $\frac{\log_4 256}{\log_4 16} = \frac{4}{2} = 2$ because $4^4 = 256$ and $4^2 = 16$.
- Both sides equal 2.

4) Plug $b = 100$, $x = 0.000001$, and $a = 10$ into the change of base formula.

$$\log_{100} 0.000001 = \frac{\log_a x}{\log_a b} = \frac{\log_{10} 0.000001}{\log_{10} 100}$$

- The left side equals $\log_{100} 0.000001 = \boxed{-3}$ because $100^{-3} = \frac{1}{100^3} = \frac{1}{1,000,000} = 0.000001$.
- The right side equals $\frac{\log_{10} 0.000001}{\log_{10} 100} = \frac{-6}{2} = -3$ because $10^{-6} = \frac{1}{10^6} = \frac{1}{1,000,000} = 0.000001$ and $10^2 = 100$.
- Both sides equal -3.

5) Plug $b = 2$, $x = 4096$, and $a = 16$ into the change of base formula.

$$\log_2 4096 = \frac{\log_a x}{\log_a b} = \frac{\log_{16} 4096}{\log_{16} 2}$$

- The left side equals $4096 = \boxed{12}$ because $2^{12} = 4096$.
- The right side equals $\frac{\log_{16} 4096}{\log_{16} 2} = \frac{3}{1/4} = 12$ because $16^3 = 4096$ and $16^{1/4} = \sqrt[4]{16} = 2$. Also note that $\frac{3}{1/4} = \frac{3}{1} \div \frac{1}{4} = \frac{3}{1} \times \frac{4}{1} = \frac{12}{1} = 12$ (or $\frac{3}{1/4} = \frac{3}{0.25} = 12$).
- Both sides equal 12.

6) Plug $b = 9$, $x = 27$, and $a = 3$ into the change of base formula.

$$\log_9 27 = \frac{\log_a x}{\log_a b} = \frac{\log_3 27}{\log_3 9}$$

- The left side equals $\log_9 27 = \boxed{\frac{3}{2}} = \boxed{1.5}$ because $9^{3/2} = \left(\sqrt{9}\right)^3 = 3^3 = 27$.
- The right side equals $\frac{\log_3 27}{\log_3 9} = \frac{3}{2} = 1.5$ because $3^3 = 27$ and $3^2 = 9$.
- Both sides equal $\frac{3}{2} = 1.5$.

Answer Key

Exercise Set 4.3

1) $\log_2 9 \approx \frac{\ln 9}{\ln 2}\left(\text{or } \frac{\log_{10} 9}{\log_{10} 2}\right) \approx 3.17$

Note: You can check the answers by raising the base to the power. Example: $2^{3.17} \approx 9$

2) $\log_{11} 3 \approx \frac{\ln 3}{\ln 11}\left(\text{or } \frac{\log_{10} 3}{\log_{10} 11}\right) \approx 0.458$

3) $\log_4 0.71 \approx \frac{\ln 0.71}{\ln 4}\left(\text{or } \frac{\log_{10} 0.71}{\log_{10} 4}\right) \approx -0.247$

4) $\log_3 \left(\frac{1}{2}\right) \approx \frac{\ln(1/2)}{\ln 3}\left(\text{or } \frac{\log_{10}(1/2)}{\log_{10} 3}\right) \approx -0.631$

5) $\log_7 42 \approx \frac{\ln 42}{\ln 7}\left(\text{or } \frac{\log_{10} 42}{\log_{10} 7}\right) \approx 1.92$

6) $\log_6 813 \approx \frac{\ln 813}{\ln 6}\left(\text{or } \frac{\log_{10} 813}{\log_{10} 6}\right) \approx 3.74$

7) $\log_5 37.5 \approx \frac{\ln 37.5}{\ln 5}\left(\text{or } \frac{\log_{10} 37.5}{\log_{10} 5}\right) \approx 2.25$

8) $\log_{12} 2.107 \approx \frac{\ln 2.107}{\ln 12}\left(\text{or } \frac{\log_{10} 2.107}{\log_{10} 12}\right) \approx 0.300$ (round 0.2999 up to 0.300)

9) $\log_{63} 2 \approx \frac{\ln 2}{\ln 63}\left(\text{or } \frac{\log_{10} 2}{\log_{10} 63}\right) \approx 0.167$

10) $\log_9 \left(\frac{3}{4}\right) \approx \frac{\ln(3/4)}{\ln 9}\left(\text{or } \frac{\log_{10}(3/4)}{\log_{10} 9}\right) \approx -0.131$

11) $\log_2 0.007 \approx \frac{\ln 0.007}{\ln 2}\left(\text{or } \frac{\log_{10} 0.007}{\log_{10} 2}\right) \approx -7.16$

12) $\log_8 1{,}000{,}000 \approx \frac{\ln 1{,}000{,}000}{\ln 8}\left(\text{or } \frac{\log_{10} 1{,}000{,}000}{\log_{10} 8}\right) \approx 6.64$

Exercise Set 4.4

1) $\log_3(9.8^5) \approx \frac{\ln(9.8^5)}{\ln 3}\left(\text{or } \frac{\log_{10}(9.8^5)}{\log_{10} 3}\right) \approx 10.4$ Note: $9.8^5 \approx 90{,}392$

2) $\log_7 e \approx \frac{\ln e}{\ln 7}\left(\text{or } \frac{\log_{10} e}{\log_{10} 7}\right) \approx 0.514$ Note: $\ln e = 1$ whereas $\log_{10} e \approx 0.434$

Note: We kept all of the digits throughout the calculation and only rounded the final answer. Although we wrote $\log_{10} e \approx 0.434$ above, we used several digits (not just these 3) for the calculations. The same is true for the solutions of the other exercises.

3) $\log_{15}\left(\frac{42^7}{8^3}\right) \approx \frac{\ln\left(\frac{42^7}{8^3}\right)}{\ln 15}\left(\text{or } \frac{\log_{10}\left(\frac{42^7}{8^3}\right)}{\log_{10} 15}\right) \approx 7.36$ Note: $\frac{42^7}{8^3} \approx 4.50 \times 10^8$

4) $\log_3 \sqrt{375} \approx \frac{\ln \sqrt{375}}{\ln 3}$ (or $\frac{\log_{10} \sqrt{375}}{\log_{10} 3}$) ≈ 2.70 Notes: $\sqrt{375} \approx 19.4$ and 2.697 rounds up to 2.70

5) $\sqrt{\log_6 248} \approx \sqrt{\frac{\ln 248}{\ln 6}}$ (or $\sqrt{\frac{\log_{10} 248}{\log_{10} 6}}$) ≈ 1.75 Note: $\log_6 248 \approx 3.08$

Note: Take the square root after finding the logarithms.

6) $\log_5 \left(\frac{\pi^2}{3}\right) \approx \frac{\ln\left(\frac{\pi^2}{3}\right)}{\ln 5}$ (or $\frac{\log_{10}\left(\frac{\pi^2}{3}\right)}{\log_{10} 5}$) ≈ 0.740 Notes: $\frac{\pi^2}{3} \approx 3.29$ and 0.7399 rounds up to 0.740

7) $\log_8 [6^{(5^3)}] \approx \frac{\ln[6^{(5^3)}]}{\ln 8}$ (or $\frac{\log_{10}[6^{(5^3)}]}{\log_{10} 8}$) ≈ 108 Note: $6^{(5^3)} = 6^{125} \approx 1.86 \times 10^{97}$

8) $\sqrt{\log_4 (39^2)} \approx \sqrt{\frac{\ln(39^2)}{\ln 4}}$ (or $\sqrt{\frac{\log_{10}(39^2)}{\log_{10} 4}}$) ≈ 2.30 Notes: $\ln(39^2) \approx 7.33$ and 2.299 rounds up to 2.3. Take the square root after finding the logarithms.

9) $\log_5 \left(\frac{1}{e^{16}}\right) \approx \frac{\ln\left(\frac{1}{e^{16}}\right)}{\ln 5}$ (or $\frac{\log_{10}\left(\frac{1}{e^{16}}\right)}{\log_{10} 5}$) ≈ -9.94 Note: $\ln\left(\frac{1}{e^{16}}\right) = \ln(e^{-16}) = -16$

10) First find $\log_3 6 \approx \frac{\ln 6}{\ln 3}$ (or $\frac{\log_{10} 6}{\log_{10} 3}$) ≈ 1.63 (intermediate answer)

$\log_9 (\log_3 6) \approx \frac{\ln(\log_3 6)}{\ln 9}$ (or $\frac{\log_{10}(\log_3 6)}{\log_{10} 9}$) ≈ 0.223

11) $\log_6 (\sin 72°) \approx \frac{\ln(\sin 72°)}{\ln 6}$ (or $\frac{\log_{10}(\sin 72°)}{\log_{10} 6}$) ≈ -0.0280 Note: $\sin 72° \approx 0.951$

Note: First make sure that your calculator is in degrees mode. If it is in radians mode and you use 72 degrees, you won't get the right answer.

12) $\log_2 (50!) \approx \frac{\ln(50!)}{\ln 2}$ (or $\frac{\log_{10}(50!)}{\log_{10} 2}$) ≈ 214 Notes: $\ln(50!) \approx 148$ and $50! \approx 3.04 \times 10^{64}$

Note that 50! is a factorial. It means 50 times 49 times 48 times 47 etc. until you get to 1. You'll need a calculator that has a factorial function (perhaps accessible through a PRB button or a menu). Alternatively, it's possible to determine 50! by writing code (in Python, for example) or using a spreadsheet program (like Excel).

Answer Key

Chapter 5 Exponentials

Exercise Set 5.1

1) $e^2 \approx 7.39$
2) $e^{-2} \approx 0.135$
3) $e^{1/4} \approx 1.28$
4) $e^{-1/4} \approx 0.779$
5) $e^{\sqrt{7}} \approx 14.1$
6) $e^e \approx 15.2$
7) $e^\pi \approx 23.1$
8) $e^{42} \approx 1.74 \times 10^{18}$
9) $e^{-75} \approx 2.68 \times 10^{-33}$
10) $\sqrt{e} \approx 1.65$
11) $2^e \approx 6.58$
12) $2^{-e} \approx 0.152$
13) $\ln(2e) \approx 1.69$
14) $\ln\left(\frac{e}{2}\right) \approx 0.307$
15) $\sqrt{e^2 - 1} \approx 2.53$
16) $\frac{1+e}{1-e} \approx -2.16$
17) $e^{\sin 60°} \approx 2.38$
18) $e^{\sin 300°} \approx 0.421$

Note: Make sure that your calculator is in degrees mode.

Exercise Set 5.2

1) Add 12 to both sides: $8e^x = 72$
Divide by 9 on both sides: $e^x = 9$
Take the natural logarithm of both sides: $x = \ln 9 \approx \boxed{2.197}$
Check: $8e^x - 12 \approx 8e^{2.197} - 12 \approx 60$
Note: If you only keep 4 significant figures, your answers may be appear to be slightly off when you check them. If you keep additional digits throughout your calculations, your answers will match more precisely.

2) Subtract 28 from both sides: $-2e^x = -24$
Divide by -2 on both sides: $e^x = 12$
Take the natural logarithm of both sides: $x = \ln 12 \approx \boxed{2.485}$
Check: $28 - 2e^x \approx 28 - 2e^{2.485} \approx 4$

3) Subtract 18 from both sides: $9e^{-x} = 36$
Divide by 9 on both sides: $e^{-x} = 4$
Take the natural logarithm of both sides: $-x = \ln 4$
Multiply by -1 on both sides: $x = -\ln 4 \approx \boxed{-1.386}$
Check: $9e^{-x} + 18 \approx 9e^{-(-1.386)} + 18 = 9e^{1.386} + 18 \approx 54$

Logarithms and Exponentials Essential Skills Practice Workbook with Answers

4) Take the natural logarithm of both sides: $\ln(e^{3x+2}) = \ln 4$
Apply the cancellation equation: $3x + 2 = \ln 4$
Subtract 2 from both sides: $3x = \ln 4 - 2$
Divide by 3 on both sides: $x = \frac{\ln 4 - 2}{3} \approx \boxed{-0.2046}$
Check: $e^{3x+2} \approx e^{3(-0.2046)+2} \approx 4$

5) Recall that $e^m e^n = e^{m+n}$ from Chapter 1: $e^{4x+3} = 25$
Take the natural logarithm of both sides: $\ln(e^{4x+3}) = \ln 25$
Apply the cancellation equation: $4x + 3 = \ln 25$
Subtract 3 from both sides: $4x = \ln 25 - 3$
Divide by 4 on both sides: $x = \frac{\ln 25 - 3}{4} \approx \boxed{0.05472}$
Check: $e^{4x} e^3 \approx e^{4(0.05472)} e^3 \approx 25$

6) Recall that $e^m e^n = e^{m+n}$ from Chapter 1: $e^{2x+x} = \frac{1}{100}$
Combine like terms in the exponent: $e^{3x} = \frac{1}{100}$
Take the natural logarithm of both sides: $3x = \ln\left(\frac{1}{100}\right)$
Divide by 3 on both sides: $x = \frac{1}{3}\ln\left(\frac{1}{100}\right) \approx \boxed{-1.535}$
Check: $e^{2x} e^x \approx e^{2(-1.535)} e^{-1.535} \approx 0.01 = \frac{1}{100}$

7) Multiply by $e^x - 2$ on both sides: $12 = 3(e^x - 2)$
(You should recognize the concept of "cross multiplying" from algebra.)
Divide by 4 on both sides: $4 = e^x - 2$
Add 2 from both sides: $6 = e^x$
Take the natural logarithm of both sides: $\ln 6 = x \approx \boxed{1.792}$
Check: $\frac{12}{e^x - 2} \approx \frac{12}{e^{1.792} - 2} \approx \frac{12}{4.001} \approx 3$

8) Define $u = e^x$ such that $u^2 = (e^x)^2 = e^{2x}$. The equation becomes: $u^2 + 21 = 10u$
This is a quadratic equation. Put it in standard form: $u^2 - 10u + 21 = 0$
Either use the quadratic formula or factor it as: $(u - 3)(u - 7) = 0$
These are not the final answers (since this is for u, not x): $u = 3$ or $u = 7$
Since $u = e^x$, this gives $e^x = 3$ or $e^x = 7$
Take the natural logarithm of both sides: $x = \ln 3 \approx \boxed{1.099}$ or $x = \ln 7 \approx \boxed{1.946}$
Check: $e^{2x} + 21 \approx e^{2(1.099)} + 21 \approx 30.01$ agrees with $10e^x \approx 10e^{1.099} \approx 30.01$ and
$e^{2x} + 21 \approx e^{2(1.946)} + 21 \approx 70.01$ agrees with $10e^x \approx 10e^{1.946} \approx 70.01$

Answer Key

9) Let $u = e^x$. The equation becomes $e^u = 1000$

Take the natural logarithm of both sides to find that $u = \ln 1000 \approx 6.908$ (this is not the final answer because this is u, not x)

Now we know that $6.908 \approx e^x$. Take the natural logarithm of both sides: $\ln 6.908 \approx x$

The final answer is: $x \approx \ln 6.908 \approx \boxed{1.933}$

Alternate solution: Take the natural logarithm of both sides: $\ln\left[e^{(e^x)}\right] = \ln 1000$

Apply the cancellation equation: $e^x = \ln 1000$

Take the natural logarithm of both sides again: $x = \ln(\ln 1000) \approx 1.933$

Check: $e^{(e^x)} = e^{(e^{1.933})} \approx e^{6.910} \approx 1000$

10) Take the natural logarithm of both sides: $\cos x = \ln\left(\frac{1}{2}\right)$

Take the inverse cosine of both sides: $x = \cos^{-1}\left[\ln\left(\frac{1}{2}\right)\right] \approx \cos^{-1}(-0.6931)$

The Quadrant II answer is: $x \approx \boxed{2.337}$ radians or $\boxed{133.9°}$.

Check: $e^{\cos x} \approx e^{\cos(2.337 \text{ rad})} \approx e^{\cos(133.9°)} = \frac{1}{2}$

Exercise Set 5.3

1) $4^{7.2} \approx 2.16 \times 10^4$

2) $7^{2/3} \approx 3.66$

3) $25^{-6.9} \approx 2.26 \times 10^{-10}$

4) $6^{-3/4} \approx 0.261$

5) $5^{\sqrt{3}} \approx 16.2$

6) $9^\pi \approx 995$

7) $\pi^\pi \approx 36.5$

8) $8^{\sqrt{5}} \approx 105$

9) $\left(\frac{4}{7}\right)^{3/5} \approx 0.715$

10) $\sqrt{e^\pi} \approx 4.81$

11) $\sqrt{7}^{\sqrt{11}} \approx 25.2$

12) $\sqrt{7^{11}} \approx 4.45 \times 10^4$

Exercise Set 5.4

1) Add 18 to both sides: $3^x = 54$

Take a base-3 logarithm of both sides: $x = \log_3 54$

Use the change of base formula: $x = \frac{\ln 54}{\ln 3}$ or $\frac{\log_{10} 54}{\log_{10} 3} \approx \boxed{3.631}$

Check: $3^x - 18 \approx 3^{3.631} - 18 \approx 36$

Logarithms and Exponentials Essential Skills Practice Workbook with Answers

2) Subtract 25 from both sides: $5^{-x} = 125$

Take a base-5 logarithm of both sides: $-x = \log_5 125$

Multiply by -1 on both sides: $x = -\log_5 125$

You should be able to do this calculation in your head: $x = -3$ (exactly)

Use the change of base formula: $x = -\frac{\ln 125}{\ln 5}$ or $-\frac{\log_{10} 125}{\log_{10} 5} = \boxed{-3}$

Check: $5^{-x} + 25 = 5^{-(-3)} + 25 = 5^3 + 25 = 125 + 25 = 150$

Note: $x = -3$ is negative such that $-x = -(-3) = 3$ is positive.

3) Take a base-8 logarithm of both sides: $5x - 2 = \log_8\left(\frac{2}{3}\right)$

(If it helps, let $u = 5x - 2$. After you find u, then replace it with $5x - 2$.)

Add 2 to both sides: $5x = 2 + \log_8\left(\frac{2}{3}\right)$

Divide by 5 on both sides: $x = \frac{2}{5} + \frac{1}{5}\log_8\left(\frac{2}{3}\right)$

Use the change of base formula: $x = \frac{2}{5} + \frac{1}{5}\frac{\ln(2/3)}{\ln 8}$ or $\frac{2}{5} + \frac{1}{5}\frac{\log_{10}(2/3)}{\log_{10} 8} \approx \boxed{0.3610}$

Note: After finding $\frac{\ln(2/3)}{\ln 8}$, divide by 5 and then add 2/5.

Check: $8^{5x-2} \approx 8^{5(0.3610)-2} \approx 0.67 \approx \frac{2}{3}$

4) Apply the rule $\frac{b^m}{b^n} = b^{m-n}$ (Chapter 1) to write: $7^{5x} = \frac{3}{50}$

Take a base-7 logarithm of both sides: $5x = \log_7\left(\frac{3}{50}\right)$

Divide by 5 on both sides: $x = \frac{1}{5}\log_7\left(\frac{3}{50}\right)$

Use the change of base formula: $x = \frac{1}{5}\frac{\ln(3/50)}{\ln 7}$ or $\frac{1}{5}\frac{\log_{10}(3/50)}{\log_{10} 7} \approx \boxed{-0.2892}$

Note: Remember to divide by 5 after you use the change of base formula.

Check: $\frac{7^{9x}}{7^{4x}} \approx \frac{7^{9(-0.2892)}}{7^{4(-0.2892)}} \approx 0.06 = \frac{3}{50}$

5) Note that $4^x = (2^2)^x = 2^{2x}$ and $8^x = (2^3)^x = 2^{3x}$ because $(b^m)^n = b^{mn}$

The given equation is equivalent to: $2^x 2^{2x} 2^{3x} = 175$

Apply the rule $b^m b^n = b^{m+n}$ (Chapter 1) to write: $2^{6x} = 175$

Take a base-2 logarithm of both sides: $6x = \log_2 175$

Divide by 6 on both sides: $x = \frac{1}{6}\log_2 175$

Answer Key

Use the change of base formula: $x = \frac{1}{6}\frac{\ln 175}{\ln 2}$ or $\frac{1}{6}\frac{\log_{10} 175}{\log_{10} 2} \approx \boxed{1.242}$

Note: Remember to divide by 6 after you use the change of base formula.

Check: $2^x 4^x 8^x \approx 2^{1.242} 4^{1.242} 8^{1.242} \approx 175$

6) Note that $36^x = (6^2)^x = 6^{2x}$ because $(b^m)^n = b^{mn}$

The given equation is equivalent to: $6^x 6^{2x} = 1296$

Apply the rule $b^m b^n = b^{m+n}$ (Chapter 1) to write: $6^{3x} = 1296$

You could solve this problem like we solved Problem 5.

One alternative is to recognize that $1296 = 6^4$ such that $6^{3x} = 6^4$

In order for $6^{3x} = 6^4$ to be true, it must be true that $3x = 4$

Divide by 3 on both sides to get $x = \boxed{\frac{4}{3}}$ (exactly, which rounds to $\boxed{1.333}$)

Check: $6^x 36^x \approx 6^{1.333} 36^{1.333} \approx 1296$ (if you use 4/3 instead of 1.333, it will be exact)

7) Multiply by 2 and multiply by $6^x + 1$ on both sides to get: $48 = 6^x + 1$

(You should recognize the concept of "cross multiplying" from algebra.)

Subtract 1 from both sides to get $47 = 6^x$

Take a base-6 logarithm of both sides: $\log_6 47 = x$

Use the change of base formula: $x = \frac{\ln 47}{\ln 6}$ or $\frac{\log_{10} 47}{\log_{10} 6} \approx \boxed{2.149}$

Check: $\frac{24}{6^x+1} \approx \frac{24}{6^{2.149}+1} \approx 0.5 = \frac{1}{2}$

8) Let $u = 2^x$. The equation becomes $3^u = 500$

Take the base-3 logarithm of both sides: $u = \log_3 500$ (not the final answer)

Use the change of base formula: $u = \frac{\ln 500}{\ln 3}$ or $\frac{\log_{10} 500}{\log_{10} 3} \approx 5.657$ (not the final answer)

(This is not the final answer because this is u, not x)

Since $u = 2^x$ and $u \approx 5.657$, now we know that $5.657 \approx 2^x$

Take the base-2 logarithm of both sides: $\log_2 5.657 \approx x$

Use the change of base formula: $x = \frac{\ln 5.657}{\ln 2}$ or $\frac{\log_{10} 5.657}{\log_{10} 2} \approx 2.500$ (final answer)

Alternate solution: Take the base-3 logarithm of both sides: $2^x = \log_3 500$

Take the base-2 logarithm of both sides again: $x = \log_2(\log_3 500)$

Use the change of base formula twice: $x = \frac{\ln\left(\frac{\ln 500}{\ln 3}\right)}{\ln 2} \approx \frac{\ln 5.657}{\ln 2} \approx \boxed{2.500}$

Check: $3^{(2^x)} \approx 3^{(2^{2.500})} \approx 3^{5.657} \approx 500$

9) Define $u = 5^x$ such that $u^2 = (5^x)^2 = 5^{2x}$. The equation becomes: $u^2 + 24 = 11u$
This is a quadratic equation. Put it in standard form: $u^2 - 11u + 24 = 0$
Either use the quadratic formula or factor it as: $(u - 3)(u - 8) = 0$
These are not the final answers (since this is for u, not x): $u = 3$ or $u = 8$
Since $u = 5^x$, this gives $5^x = 3$ or $5^x = 8$
Take the base-5 logarithm of both sides: $x = \log_5 3$ or $x = \log_5 8$
Use the change of base formula: $x = \frac{\ln 3}{\ln 5}$ or $\frac{\log_{10} 3}{\log_{10} 5} \approx \boxed{0.6826}$ or $x = \frac{\ln 8}{\ln 5}$ or $\frac{\log_{10} 8}{\log_{10} 5} \approx \boxed{1.292}$
Check: $5^{2x} + 24 \approx 5^{2(0.6826)} + 24 \approx 33$ agrees with $11(5^x) \approx 11(5^{0.6826}) \approx 33$ and
$5^{2x} + 24 \approx 5^{2(1.292)} + 24 \approx 88$ agrees with $11(5^x) \approx 11(5^{1.292}) \approx 88$

10) Multiply by 5 and 2 on both sides (equivalent to multiplying by 10 on both sides)
to get: $2(9^x) + 2(1) = 5(3^x)$, which simplifies to $2(9^x) + 2 = 5(3^x)$
(You should recognize the concept of "cross multiplying" from algebra.)
Note that $9^x = (3^2)^x = 3^{2x}$ because $(b^m)^n = b^{mn}$
The previous equation becomes: $2(3^{2x}) + 2 = 5(3^x)$
Define $u = 3^x$ such that $u^2 = (3^x)^2 = 3^{2x}$. The equation becomes: $2u^2 + 2 = 5u$
This is a quadratic equation. Put it in standard form: $2u^2 - 5u + 2 = 0$
Either use the quadratic formula or factor it as: $(2u - 1)(u - 2) = 0$
These are not the final answers (since this is for u, not x): $u = \frac{1}{2}$ or $u = 2$
Since $u = 3^x$, this gives $3^x = \frac{1}{2}$ or $3^x = 2$
Take the base-3 logarithm of both sides: $x = \log_3\left(\frac{1}{2}\right)$ or $x = \log_3 2$
Use the change of base formula: $x = \frac{\ln(1/2)}{\ln 3}$ or $\frac{\log_{10}(1/2)}{\log_{10} 3} \approx \boxed{-0.6309}$ or $x = \frac{\ln 2}{\ln 3}$ or $\frac{\log_{10} 2}{\log_{10} 3} \approx \boxed{0.6309}$
Check: $\frac{9^x+1}{5} \approx \frac{9^{-0.6309}+1}{5} \approx 0.25$ agrees with $\frac{3^x}{2} \approx \frac{3^{-0.6309}}{2} \approx 0.25$ and
$\frac{9^x+1}{5} \approx \frac{9^{0.6309}+1}{5} \approx 1$ agrees with $\frac{3^x}{2} \approx \frac{3^{0.6309}}{2} \approx 1$

Chapter 6 Hyperbolic Functions

Exercise Set 6.1

1) $\sinh 0 = \frac{e^0 - e^0}{2} = \frac{1-1}{2} = 0$ (exactly)

2) $\cosh 0 = \frac{e^0 + e^0}{2} = \frac{1+1}{2} = \frac{2}{2} = 1$ (exactly)

3) $\sinh 1 = \frac{e^1 - e^{-1}}{2} \approx 1.175$

4) $\cosh 1 = \frac{e^1 + e^{-1}}{2} \approx 1.543$

5) $\sinh 2 = \frac{e^2 - e^{-2}}{2} \approx 3.627$

6) $\cosh 2 = \frac{e^2 + e^{-2}}{2} \approx 3.762$

7) $\sinh\left(\frac{1}{2}\right) = \frac{e^{1/2} - e^{-1/2}}{2} \approx 0.5211$

8) $\cosh\left(\frac{1}{2}\right) = \frac{e^{1/2} + e^{-1/2}}{2} \approx 1.128$

9) $\sinh(-1) = \frac{e^{-1} - e^{-(-1)}}{2} = \frac{e^{-1} - e^1}{2} \approx -1.175$

10) $\cosh(-1) = \frac{e^{-1} + e^{-(-1)}}{2} = \frac{e^{-1} + e^1}{2} \approx 1.543$

11) $\sinh(-2) = \frac{e^{-2} - e^{-(-2)}}{2} = \frac{e^{-2} - e^2}{2} \approx -3.627$

12) $\cosh(-2) = \frac{e^{-2} + e^{-(-2)}}{2} = \frac{e^{-2} + e^2}{2} \approx 3.762$

13) $\sinh(\ln 2) = \frac{e^{\ln 2} - e^{-\ln 2}}{2} = \frac{e^{\ln 2} - e^{\ln(1/2)}}{2} = \frac{2 - \frac{1}{2}}{2} = \frac{\frac{4}{2} - \frac{1}{2}}{2} = \frac{3/2}{2} = \frac{3}{2} \div \frac{2}{1} = \frac{3}{2} \times \frac{1}{2} = \frac{3}{4} = 0.75$

Note: $-\ln 2 = \ln\left(\frac{1}{2}\right)$ according to the rule $-\ln y = \ln\left(\frac{1}{y}\right)$ from Chapter 3.

Note: $e^{\ln 2} = 2$ and $e^{\ln(1/2)} = \frac{1}{2}$ according to the rule $e^{\ln y} = y$ from Chapter 3.

Notes: $2 - \frac{1}{2} = \frac{4}{2} - \frac{1}{2} = \frac{3}{2}$ and $\frac{3}{2} \div 2 = \frac{3}{2} \div \frac{2}{1} = \frac{3}{2} \times \frac{1}{2} = \frac{3}{4}$.

14) $\cosh(\ln 2) = \frac{e^{\ln 2} + e^{-\ln 2}}{2} = \frac{e^{\ln 2} + e^{\ln(1/2)}}{2} = \frac{2 + \frac{1}{2}}{2} = \frac{\frac{4}{2} + \frac{1}{2}}{2} = \frac{5/2}{2} = \frac{5}{2} \div \frac{2}{1} = \frac{5}{2} \times \frac{1}{2} = \frac{5}{4} = 1.25$

15) $\sinh\sqrt{2} = \frac{e^{\sqrt{2}} - e^{-\sqrt{2}}}{2} \approx 1.935$

16) $\cosh\sqrt{2} = \frac{e^{\sqrt{2}} + e^{-\sqrt{2}}}{2} \approx 2.178$

Logarithms and Exponentials Essential Skills Practice Workbook with Answers

Exercise Set 6.2

1) $\cosh x - \sinh x = \dfrac{e^x + e^{-x}}{2} - \dfrac{e^x - e^{-x}}{2} = \dfrac{e^x + e^{-x} - (e^x - e^{-x})}{2} = \dfrac{e^x + e^{-x} - e^x + e^{-x}}{2} = \dfrac{2e^{-x}}{2} = e^{-x}$

2) $\sinh(-x) = \dfrac{e^{-x} - e^{-(-x)}}{2} = \dfrac{e^{-x} - e^x}{2} = \dfrac{-e^x + e^{-x}}{2} = \dfrac{-(e^x - e^{-x})}{2} = -\sinh x$

3) $\cosh(-x) = \dfrac{e^{-x} + e^{-(-x)}}{2} = \dfrac{e^{-x} + e^x}{2} = \dfrac{e^x + e^{-x}}{2} = \cosh x$

4) $\sinh x \cosh y + \cosh x \sinh y = \dfrac{e^x - e^{-x}}{2} \dfrac{e^y + e^{-y}}{2} + \dfrac{e^x + e^{-x}}{2} \dfrac{e^y - e^{-y}}{2}$

$= \dfrac{e^x e^y + e^x e^{-y} - e^{-x} e^y - e^{-x} e^{-y}}{4} + \dfrac{e^x e^y - e^x e^{-y} + e^{-x} e^y - e^{-x} e^{-y}}{4}$

$= \dfrac{e^x e^y + e^x e^{-y} - e^{-x} e^y - e^{-x} e^{-y} + e^x e^y - e^x e^{-y} + e^{-x} e^y - e^{-x} e^{-y}}{4}$

$= \dfrac{2 e^x e^y - 2 e^{-x} e^{-y}}{4} = \dfrac{2 e^{x+y} - 2 e^{-x-y}}{4} = \dfrac{2 e^{x+y} - 2 e^{-(x+y)}}{4} = \dfrac{e^{x+y} - e^{-(x+y)}}{2} = \sinh(x+y)$

5) $\cosh x \cosh y + \sinh x \sinh y = \dfrac{e^x + e^{-x}}{2} \dfrac{e^y + e^{-y}}{2} + \dfrac{e^x - e^{-x}}{2} \dfrac{e^y - e^{-y}}{2}$

$= \dfrac{e^x e^y + e^x e^{-y} + e^{-x} e^y + e^{-x} e^{-y}}{4} + \dfrac{e^x e^y - e^x e^{-y} - e^{-x} e^y + e^{-x} e^{-y}}{4}$

$= \dfrac{e^x e^y + e^x e^{-y} + e^{-x} e^y + e^{-x} e^{-y} + e^x e^y - e^x e^{-y} - e^{-x} e^y + e^{-x} e^{-y}}{4}$

$= \dfrac{2 e^x e^y + 2 e^{-x} e^{-y}}{4} = \dfrac{2 e^{x+y} + 2 e^{-x-y}}{4} = \dfrac{2 e^{x+y} + 2 e^{-(x+y)}}{4} = \dfrac{e^{x+y} + e^{-(x+y)}}{2} = \cosh(x+y)$

6) $2 \sinh x \cosh x = 2 \dfrac{e^x - e^{-x}}{2} \dfrac{e^x + e^{-x}}{2} = 2 \dfrac{e^x e^x + e^x e^{-x} - e^{-x} e^x - e^{-x} e^{-x}}{4} = \dfrac{e^{2x} + e^0 - e^0 - e^{-2x}}{2}$

$= \dfrac{e^{2x} - e^{-2x}}{2} = \sinh(2x)$ Note: Or set $y = x$ in $\sinh x \cosh y + \cosh x \sinh y = \sinh(x+y)$.

7) $\cosh^2 x + \sinh^2 x = \left(\dfrac{e^x + e^{-x}}{2}\right)^2 + \left(\dfrac{e^x - e^{-x}}{2}\right)^2 = \dfrac{e^{2x} + 2 e^x e^{-x} + e^{-2x}}{4} + \dfrac{e^{2x} - 2 e^x e^{-x} + e^{-2x}}{4}$

$= \dfrac{e^{2x} + 2 + e^{-2x}}{4} + \dfrac{e^{2x} - 2 + e^{-2x}}{4} = \dfrac{e^{2x} + 2 + e^{-2x} + e^{2x} - 2 + e^{-2x}}{4} = \dfrac{2 e^{2x} + 2 e^{-2x}}{4} = \dfrac{e^{2x} + e^{-2x}}{2} = \cosh(2x)$

Note: $e^x e^{-x} = e^{x-x} = e^0 = 1$ Note: Or set $y = x$ in $\cosh x \cosh y + \sinh x \sinh y = \cosh(x+y)$.

8) $\cosh^2 x - \sinh^2 x = \left(\dfrac{e^x + e^{-x}}{2}\right)^2 - \left(\dfrac{e^x - e^{-x}}{2}\right)^2 = \dfrac{e^{2x} + 2 e^x e^{-x} + e^{-2x}}{4} - \dfrac{e^{2x} - 2 e^x e^{-x} + e^{-2x}}{4}$

$= \dfrac{e^{2x} + 2 + e^{-2x}}{4} - \dfrac{e^{2x} - 2 + e^{-2x}}{4} = \dfrac{e^{2x} + 2 + e^{-2x} - (e^{2x} - 2 + e^{-2x})}{4} = \dfrac{e^{2x} + 2 + e^{-2x} - e^{2x} + 2 - e^{-2x}}{4} = \dfrac{2+2}{4} = \dfrac{4}{4} = 1$

9) $(\cosh x + \sinh x)^n = \left(\dfrac{e^x + e^{-x}}{2} + \dfrac{e^x - e^{-x}}{2}\right)^n = \left(\dfrac{e^x + e^{-x} + e^x - e^{-x}}{2}\right)^n = \left(\dfrac{2e^x}{2}\right)^n = e^{nx}$

$\cosh(nx) + \sinh(nx) = \dfrac{e^{nx} + e^{-nx}}{2} + \dfrac{e^{nx} - e^{-nx}}{2} = \dfrac{e^{nx} + e^{-nx} + e^{nx} - e^{-nx}}{2} = \dfrac{2 e^{nx}}{2} = e^{nx}$

$(\cosh x + \sinh x)^n = \cosh(nx) + \sinh(nx)$ since each side equals e^{nx}

Exercise Set 6.3

1) $\tanh 0 = \dfrac{\sinh 0}{\cosh 0} = \dfrac{e^0 - e^0}{e^0 + e^0} = \dfrac{1-1}{1+1} = \dfrac{0}{2} = 0$ (exactly)

2) $\text{sech } 0 = \dfrac{1}{\cosh 0} = \dfrac{2}{e^0 + e^0} = \dfrac{2}{1+1} = \dfrac{2}{2} = 1$ (exactly)

3) $\tanh 1 = \dfrac{\sinh 1}{\cosh 1} = \dfrac{e^1 - e^{-1}}{e^1 + e^{-1}} \approx 0.7616$

4) $\coth 1 = \dfrac{\cosh 1}{\sinh 1} = \dfrac{e^1 + e^{-1}}{e^1 - e^{-1}} \approx 1.313$

5) $\text{sech } 2 = \dfrac{1}{\cosh 2} = \dfrac{2}{e^2 + e^{-2}} \approx 0.2658$

6) $\text{csch } 2 = \dfrac{1}{\sinh 2} = \dfrac{2}{e^2 - e^{-2}} \approx 0.2757$

7) $\tanh 2 = \dfrac{\sinh 2}{\cosh 2} = \dfrac{e^2 - e^{-2}}{e^2 + e^{-2}} \approx 0.9640$

8) $\coth 2 = \dfrac{\cosh 2}{\sinh 2} = \dfrac{e^2 + e^{-2}}{e^2 - e^{-2}} \approx 1.037$

9) $\text{sech}\left(\dfrac{1}{2}\right) = \dfrac{1}{\cosh(1/2)} = \dfrac{2}{e^{1/2} + e^{-1/2}} \approx 0.8868$

10) $\text{csch}\left(\dfrac{1}{2}\right) = \dfrac{1}{\sinh(1/2)} = \dfrac{2}{e^{1/2} - e^{-1/2}} \approx 1.919$

11) $\tanh\left(\dfrac{1}{2}\right) = \dfrac{\sinh(1/2)}{\cosh(1/2)} = \dfrac{e^{1/2} - e^{-1/2}}{e^{1/2} + e^{-1/2}} \approx 0.4621$

12) $\coth\left(\dfrac{1}{2}\right) = \dfrac{\cosh(1/2)}{\sinh(1/2)} = \dfrac{e^{1/2} + e^{-1/2}}{e^{1/2} - e^{-1/2}} \approx 2.164$

13) $\text{sech}(\ln 2) = \dfrac{1}{\cosh(\ln 2)} = \dfrac{2}{e^{\ln 2} + e^{-\ln 2}} = \dfrac{2}{e^{\ln 2} + e^{\ln(1/2)}} = \dfrac{2}{2 + \frac{1}{2}} = \dfrac{2}{\frac{4+1}{2}} = \dfrac{2}{5/2} = 2 \div \dfrac{5}{2} = 2 \times \dfrac{2}{5} = \dfrac{4}{5} = 0.8$

Note: $-\ln 2 = \ln\left(\dfrac{1}{2}\right)$ according to the rule $-\ln y = \ln\left(\dfrac{1}{y}\right)$ from Chapter 3.

Note: $e^{\ln 2} = 2$ and $e^{\ln(1/2)} = \dfrac{1}{2}$ according to the rule $e^{\ln y} = y$ from Chapter 3.

Notes: $2 + \dfrac{1}{2} = \dfrac{4}{2} + \dfrac{1}{2} = \dfrac{5}{2}$ and $2 \div \dfrac{5}{2} = \dfrac{2}{1} \div \dfrac{5}{2} = \dfrac{2}{1} \times \dfrac{2}{5} = \dfrac{4}{5}$.

14) $\text{csch}(\ln 2) = \dfrac{1}{\text{sech}(\ln 2)} = \dfrac{2}{e^{\ln 2} - e^{-\ln 2}} = \dfrac{2}{e^{\ln 2} - e^{\ln(1/2)}} = \dfrac{2}{2 - \frac{1}{2}} = \dfrac{2}{\frac{4-1}{2}} = \dfrac{2}{3/2} = 2 \div \dfrac{3}{2} = 2 \times \dfrac{2}{3} = \dfrac{4}{3} = 1.33$

15) $\tanh\sqrt{2} = \dfrac{\sinh\sqrt{2}}{\cosh\sqrt{2}} = \dfrac{e^{\sqrt{2}} - e^{-\sqrt{2}}}{e^{\sqrt{2}} + e^{-\sqrt{2}}} \approx 0.8884$

16) $\coth\sqrt{2} = \dfrac{\cosh\sqrt{2}}{\sinh\sqrt{2}} = \dfrac{e^{\sqrt{2}} + e^{-\sqrt{2}}}{e^{\sqrt{2}} - e^{-\sqrt{2}}} \approx 1.126$

Logarithms and Exponentials Essential Skills Practice Workbook with Answers

Exercise Set 6.4

1) $\tanh(-x) = \frac{\sinh(-x)}{\cosh(-x)} = \frac{-\sinh x}{\cosh x} = -\tanh x$ Note: See Problems 2-3 in Sec. 6.2.

2) $\text{sech}(-x) = \frac{1}{\cosh(-x)} = \frac{1}{\cosh x} = \text{sech } x$ Note: See Problem 3 in Sec. 6.2.

3) $\text{csch}(-x) = \frac{1}{\sinh(-x)} = \frac{1}{-\sinh x} = -\text{csch } x$ Note: See Problem 2 in Sec. 6.2.

4) $\coth^2 x - 1 = \frac{\cosh^2 x}{\sinh^2 x} - 1 = \frac{\cosh^2 x}{\sinh^2 x} - \frac{\sinh^2 x}{\sinh^2 x} = \frac{\cosh^2 x - \sinh^2 x}{\sinh^2 x} = \frac{1}{\sinh^2 x} = \text{csch}^2 x$

Note: See Problem 8 in Sec. 6.2.

5) $\tanh(x + y) = \frac{\sinh(x+y)}{\cosh(x+y)} = \frac{\sinh x \cosh y + \cosh x \sinh y}{\cosh x \cosh y + \sinh x \sinh y}$

$= \frac{\frac{\sinh x \cosh y}{\cosh x \cosh y} + \frac{\cosh x \sinh y}{\cosh x \cosh y}}{\frac{\cosh x \cosh y}{\cosh x \cosh y} + \frac{\sinh x \sinh y}{\cosh x \cosh y}} = \frac{(\tanh x)(1) + (1)(\tanh y)}{(1)(1) + \tanh x \tanh y} = \frac{\tanh x + \tanh y}{1 + \tanh x \tanh y}$

Note: We divided every term of the numerator and denominator by $\cosh x \cosh y$.

Note: See Problems 4-5 in Sec. 6.2.

6) $\tanh(2x) = \frac{\tanh x + \tanh x}{1 + \tanh x \tanh x} = \frac{2 \tanh x}{1 + \tanh^2 x}$ Note: Set $y = x$ in the previous answer.

7) $\tanh\left(\frac{x}{2}\right) = \frac{e^{x/2} - e^{-x/2}}{e^{x/2} + e^{-x/2}} = \frac{e^{x/2} - e^{-x/2}}{e^{x/2} + e^{-x/2}} \frac{e^{x/2} + e^{-x/2}}{e^{x/2} + e^{-x/2}} = \frac{e^x + e^{x/2}e^{-x/2} - e^{x/2}e^{x/2} - e^{-x}}{e^x + e^{x/2}e^{-x/2} + e^{x/2}e^{x/2} + e^{-x}}$

$= \frac{e^x + 1 - 1 - e^{-x}}{e^x + 1 + 1 + e^{-x}} = \frac{e^x - e^{-x}}{2 + e^x + e^{-x}} = \frac{\frac{e^x - e^{-x}}{2}}{\frac{2 + e^x + e^{-x}}{2}} = \frac{\frac{e^x - e^{-x}}{2}}{\frac{2}{2} + \frac{e^x + e^{-x}}{2}} = \frac{\frac{e^x - e^{-x}}{2}}{1 + \frac{e^x + e^{-x}}{2}} = \frac{\sinh x}{1 + \cosh x}$

Note: $e^{x/2} e^{-x/2} = e^{x/2 - x/2} = e^0 = 1$

Notes: Partway through the first line, we multiplied the numerator and denominator each by $e^{x/2} + e^{-x/2}$. Partway through the second line, we divided the numerator and denominator each by 2.

8) $\tanh(\ln x) = \frac{e^{\ln x} - e^{-\ln x}}{e^{\ln x} + e^{-\ln x}} = \frac{e^{\ln x} - e^{\ln(1/x)}}{e^{\ln x} + e^{\ln(1/x)}} = \frac{x - \frac{1}{x}}{x + \frac{1}{x}} = \frac{x\left(x - \frac{1}{x}\right)}{x\left(x + \frac{1}{x}\right)} = \frac{x^2 - \frac{x}{x}}{x^2 + \frac{x}{x}} = \frac{x^2 - 1}{x^2 + 1}$

Note: We multiplied the numerator and denominator each by x.

Note: Recall from Chapter 3 that $-\ln x = \ln\left(\frac{1}{x}\right)$ and that $e^{\ln y} = y$.

9) $\frac{1 + \tanh x}{1 - \tanh x} = \frac{1 + \frac{e^x - e^{-x}}{e^x + e^{-x}}}{1 - \frac{e^x - e^{-x}}{e^x + e^{-x}}} = \frac{\frac{e^x + e^{-x}}{e^x + e^{-x}} + \frac{e^x - e^{-x}}{e^x + e^{-x}}}{\frac{e^x + e^{-x}}{e^x + e^{-x}} - \frac{e^x - e^{-x}}{e^x + e^{-x}}} = \frac{\frac{e^x + e^{-x} + (e^x - e^{-x})}{e^x + e^{-x}}}{\frac{e^x + e^{-x} - (e^x - e^{-x})}{e^x + e^{-x}}}$ (continued on the next page)

Note that equal denominators cancel in division: $\frac{a/b}{c/b} = \frac{a}{b} \div \frac{c}{b} = \frac{a}{b} \times \frac{b}{c} = \frac{a}{c}$.

$$= \frac{e^x+e^{-x}+(e^x-e^{-x})}{e^x+e^{-x}-(e^x-e^{-x})} = \frac{e^x+e^{-x}+e^x-e^{-x}}{e^x+e^{-x}-e^x+e^{-x}} = \frac{2e^x}{2e^{-x}} = \frac{e^x}{e^{-x}} = e^{x-(-x)} = e^{x+x} = e^{2x}$$

Note: $\frac{e^x}{e^{-x}} = e^{x-(-x)} = e^{x+x} = e^{2x}$ using the rule $\frac{y^m}{y^n} = y^{m-n}$ from Sec. 1.5.

Exercise Set 6.5

1) $\cosh^{-1} 4 \approx \pm 2.063$ Check: $\cosh(\pm 2.063) = \frac{e^{2.063}+e^{-2.063}}{2} \approx 4$

2) $\sinh^{-1} 2 \approx 1.444$ Check: $\sinh 1.444 = \frac{e^{1.444}-e^{-1.444}}{2} \approx 2$

3) $\tanh^{-1} 0.5 \approx 0.5493$ Check: $\tanh 0.5493 = \frac{e^{0.5493}-e^{-0.5493}}{e^{0.5493}+e^{-0.5493}} \approx 0.5$

4) $\text{sech}^{-1} 0.5 = \cosh^{-1}\left(\frac{1}{0.5}\right) \approx \pm 1.307$ Check: $\text{sech}(\pm 1.307) = \frac{2}{e^{1.307}+e^{-1.307}} \approx 0.5$

5) $\text{csch}^{-1} 0.75 = \sinh^{-1}\left(\frac{1}{0.75}\right) \approx 1.099$ Check: $\text{csch}\, 1.099 = \frac{2}{e^{1.099}-e^{-1.099}} \approx 0.75$

6) $\coth^{-1} 10 = \tanh^{-1}\left(\frac{1}{10}\right) \approx 0.09967$ Check: $\coth 0.09967 = \frac{e^{0.09967}+e^{-0.09967}}{e^{0.09967}-e^{-0.09967}} \approx 10$

7) $\sinh^{-1}\left(-\frac{1}{4}\right) \approx -0.2475$ Check: $\sinh(-0.2475) = \frac{e^{-0.2475}-e^{-(-0.2475)}}{2} \approx -0.25 = -\frac{1}{4}$

8) $\text{sech}^{-1}\left(\frac{4}{5}\right) = \cosh^{-1}\left(\frac{5}{4}\right) \approx \pm 0.6931$ Check: $\text{sech}(\pm 0.6931) = \frac{2}{e^{0.6931}+e^{-0.6931}} \approx 0.8 = \frac{4}{5}$

9) $\tanh^{-1}\left(\frac{2}{3}\right) \approx 0.8047$ Check: $\tanh 0.8047 = \frac{e^{0.8047}-e^{-0.8047}}{e^{0.8047}+e^{-0.8047}} \approx 0.67 \approx \frac{2}{3}$

10) $\cosh^{-1}\left(\frac{5}{2}\right) \approx \pm 1.567$ Check: $\cosh(\pm 1.567) = \frac{e^{1.567}+e^{-1.567}}{2} \approx 2.5 = \frac{5}{2}$

11) $\text{csch}^{-1}\left(\frac{3}{7}\right) = \sinh^{-1}\left(\frac{7}{3}\right) \approx 1.583$ Check: $\text{csch}\, 1.583 = \frac{2}{e^{1.583}-e^{-1.583}} \approx 0.4288 \approx \frac{3}{7}$

12) $\tanh^{-1}\left(-\frac{3}{4}\right) \approx -0.9730$ Check: $\tanh(-0.9730) = \frac{e^{-0.9730}-e^{-(-0.9730)}}{e^{-0.9730}+e^{-(-0.9730)}} \approx -0.75 = -\frac{3}{4}$

13) $\coth^{-1}\sqrt{3} = \tanh^{-1}\left(\frac{1}{\sqrt{3}}\right) \approx 0.6585$ Check: $\coth 0.6585 = \frac{e^{0.6585}+e^{-0.6585}}{e^{0.6585}-e^{-0.6585}} \approx 1.732 \approx \sqrt{3}$

14) $\cosh^{-1}\sqrt{2} \approx \pm 0.8814$ Check: $\cosh(\pm 0.8814) = \frac{e^{0.8814}+e^{-0.8814}}{2} \approx 1.414 \approx \sqrt{2}$

15) $\sinh^{-1} e \approx 1.725$ Check: $\sinh 1.725 = \frac{e^{1.725}-e^{-1.725}}{2} \approx 2.717 \approx e$

16) $\text{csch}^{-1}\pi = \sinh^{-1}\left(\frac{1}{\pi}\right) \approx 0.3132$ Check: $\text{csch}\, 0.3132 = \frac{2}{e^{0.3132}-e^{-0.3132}} \approx 3.14 \approx \pi$

Note: Some books and authors use the terms arcsine, arctangent, etc. instead of inverse sine, inverse tangent, etc. Those texts would write asinh(x) instead of $\sinh^{-1}(x)$.

Logarithms and Exponentials Essential Skills Practice Workbook with Answers

Exercise Set 6.6

1) Add 2 to both sides: $5\cosh x = 8$

Divide by 5 on both sides: $\cosh x = \dfrac{8}{5}$

Take the inverse hyperbolic cosine of both sides: $x = \cosh^{-1}\left(\dfrac{8}{5}\right) \approx \boxed{\pm 1.047}$

Check: $5\cosh x - 2 = 5\cosh(\pm 1.047) - 2 \approx 8 - 2 \approx 6$

Note: Recall that even inverse functions like $\cosh^{-1}$ have two answers (with opposite signs) because $\cosh(-x) = \cosh x$. That is, $\cosh(-1.047) = \cosh 1.047$.

2) Square root both sides: $\tanh x = \pm\sqrt{\dfrac{3}{4}}$

Take the inverse hyperbolic tangent of both sides: $x = \tanh^{-1}\left(\pm\sqrt{\dfrac{3}{4}}\right) \approx \boxed{\pm 1.317}$

Check: $\tanh^2 x = \tanh^2(\pm 1.317) \approx (\pm 0.8660)^2 \approx 0.75 = \dfrac{3}{4}$

3) Take the inverse hyperbolic sine of both sides: $\sqrt{x} = \sinh^{-1} 1.5$

Square both sides: $x = (\sinh^{-1} 1.5)^2 \approx (1.195)^2 \approx \boxed{1.428}$

Check: $\sinh\sqrt{x} = \sinh\sqrt{1.428} \approx 1.5$

4) Divide by 4 on both sides: $\sinh^{-1} x = \dfrac{7}{4}$

Take the hyperbolic sine of both sides: $x = \sinh\left(\dfrac{7}{4}\right) \approx \boxed{2.790}$ (Note that this isn't an inverse. In this problem, the given equation had an inverse, not the solution.)

Check: $4\sinh^{-1} x = 4\sinh^{-1} 2.790 \approx 7$

5) Divide by hyperbolic cosine on both sides: $8\dfrac{\sinh x}{\cosh x} = 5$

Since $\dfrac{\sinh x}{\cosh x} = \tanh x$ (Sec. 6.3), this becomes: $8\tanh x = 5$

Divide by 8 on both sides: $\tanh x = \dfrac{5}{8}$

Take the inverse hyperbolic tangent of both sides: $x = \tanh^{-1}\left(\dfrac{5}{8}\right) \approx \boxed{0.7332}$

Check: $8\sinh x = 8\sinh 0.7332 \approx 6.405$ agrees with $5\cosh x = 5\cosh 0.7332 \approx 6.405$

6) Apply the identity $\cosh^2 x - \sinh^2 x = 1$ from Problem 8 of Sec. 6.2

Add $\sinh^2 x$ to both sides of $\cosh^2 x - \sinh^2 x = 1$ to get $\cosh^2 x = \sinh^2 x + 1$

Replace $\cosh^2 x$ with $\sinh^2 x + 1$ in $\sinh^2 x + \cosh^2 x = 5$

This gives $\sinh^2 x + \sinh^2 x + 1 = 5$, which simplifies to $2\sinh^2 x = 4$

Answer Key

Divide by 2 on both sides: $\sinh^2 x = 2$

Square root both sides: $\sinh x = \pm\sqrt{2}$ Note: the $\pm$ reflects that $\left(-\sqrt{2}\right)^2 = 2$ as well as $\left(\sqrt{2}\right)^2 = 2$

Take the inverse hyperbolic sine of both sides: $x = \sinh^{-1}(\pm\sqrt{2}) \approx \boxed{\pm 1.146}$

Check: $\sinh^2 x + \cosh^2 x = \sinh^2(\pm 1.146) + \cosh^2(\pm 1.146) \approx 2 + 3 = 5$

7) Apply the identity $1 - \tanh^2 x = \text{sech}^2 x$ from Example 1 of Sec. 6.4

Add $\tanh^2 x$ to both sides of $1 - \tanh^2 x = \text{sech}^2 x$ to get $1 = \text{sech}^2 x + \tanh^2 x$

Replace $\text{sech}^2 x + \tanh^2 x$ with 1 in $\text{sech}^2 x + \text{csch}^2 x + \tanh^2 x = 1.6$

This gives $1 + \text{csch}^2 x = 1.6$

Subtract 1 from both sides: $\text{csch}^2 x = 0.6$

Square root both sides: $\text{csch}\, x = \pm\sqrt{0.6}$ Note: the $\pm$ reflects that $\left(-\sqrt{0.6}\right)^2 = 0.6$ as well as $\left(\sqrt{0.6}\right)^2 = 0.6$

If you have an inverse hyperbolic cosecant function on your calculator, you can take the inverse hyperbolic cosecant on both sides. Otherwise, you can take the inverse hyperbolic sine of the reciprocal of the argument (as discussed in Sec. 6.5), as shown in a few steps below.

Since $\text{csch}\, x = \frac{1}{\sinh x}$ (Sec. 6.3), this becomes: $\frac{1}{\sinh x} = \pm\sqrt{0.6}$

Multiply by hyperbolic sine both sides: $1 = (\sinh x)(\pm\sqrt{0.6})$

Divide by $\pm\sqrt{0.6}$ on both sides: $\pm\frac{1}{\sqrt{0.6}} = \sinh x$, which is equivalent to $\sinh x = \pm\frac{1}{\sqrt{0.6}}$

Take the inverse hyperbolic sine of both sides: $x = \sinh^{-1}\left(\pm\frac{1}{\sqrt{0.6}}\right) \approx \boxed{\pm 1.073}$

Check: $\text{sech}^2 x + \text{csch}^2 x + \tanh^2 x = \text{sech}^2(\pm 1.073) + \text{csch}^2(\pm 1.073) + \tanh^2(\pm 1.073) =$
$\frac{1}{\cosh^2(\pm 1.073)} + \frac{1}{\sinh^2(\pm 1.073)} + \tanh^2(\pm 1.073) \approx (\pm 0.6123)^2 + (\pm 0.7745)^2 + (\pm 0.7906)^2 \approx$
$0.375 + 0.600 + 0.625 \approx 1.6$

8) Take the hyperbolic sine of both sides: $\cosh x = \sinh 2$

Take the inverse hyperbolic cosine of both sides: $x = \cosh^{-1}(\sinh 2) \approx \cosh^{-1}(3.627) \approx \boxed{\pm 1.962}$

Check: $\sinh^{-1}(\cosh x) = \sinh^{-1}[\cosh(\pm 1.962)] \approx \sinh^{-1}(3.627) \approx 2$

Logarithms and Exponentials Essential Skills Practice Workbook with Answers

Exercise Set 6.7

1) $\sinh^{-1} x = \ln(x + \sqrt{x^2 + 1})$

- Let $y = \sinh^{-1} x$ such that $\sinh y = x$.
- Since $\sinh y = \frac{e^y - e^{-y}}{2}$, we may rewrite the previous equation as $\frac{e^y - e^{-y}}{2} = x$.
- Multiply by 2 on both sides: $e^y - e^{-y} = 2x$.
- Subtract $2x$ from both sides: $e^y - 2x - e^{-y} = 0$.
- Multiply by e^y on both sides: $e^{2y} - 2xe^y + 1 = 0$. Note that $e^y e^y = e^{2y}$ and that $e^y e^{-y} = e^0 = 1$ because $x^m x^n = x^{m+n}$ (Chapter 1).
- $e^{2y} - 2xe^y - 1 = 0$ a quadratic equation in e^y. Apply the quadratic formula with $a = 1$, $b = -2x$, and $c = -1$ to solve for e^y:

$$e^y = \frac{-b \pm \sqrt{b^2 - 4ac}}{2a} = \frac{-(-2x) \pm \sqrt{(-2x)^2 - 4(1)(-1)}}{2(1)} = \frac{2x \pm \sqrt{4x^2 + 4}}{2}$$

- Factor the 4 out: $\sqrt{4x^2 + 4} = \sqrt{4(x^2 + 1)} = \sqrt{4}\sqrt{x^2 + 1} = 2\sqrt{x^2 + 1}$.

$$e^y = \frac{2x \pm 2\sqrt{x^2 + 1}}{2} = \frac{2x}{2} \pm \frac{2\sqrt{x^2 + 1}}{2} = x \pm \sqrt{x^2 + 1}$$

- Only the plus sign of the $\pm$ works. The minus sign would make the right-hand side negative, whereas e^y can only be positive: $e^y = x + \sqrt{x^2 + 1}$.
- Take the natural logarithm of both sides: $\ln(e^y) = \ln(x + \sqrt{x^2 + 1})$.
- Apply the cancellation equation (Chapter 3): $y = \ln(x + \sqrt{x^2 + 1})$.
- Recall that $y = \sinh^{-1} x$ from the first step: $\sinh^{-1} x = \ln(x + \sqrt{x^2 + 1})$.

2) $\tanh^{-1} x = \frac{1}{2} \ln\left(\frac{1+x}{1-x}\right)$

- Let $y = \tanh^{-1} x$ such that $\tanh y = x$.
- Since $\tanh y = \frac{e^y - e^{-y}}{e^y + e^{-y}}$, we may rewrite the previous equation as $\frac{e^y - e^{-y}}{e^y + e^{-y}} = x$.
- Multiply by $e^y + e^{-y}$ on both sides: $e^y - e^{-y} = xe^y + xe^{-y}$.
- Add e^{-y} to both sides and subtract xe^y from both sides: $e^y - xe^y = e^{-y} + xe^{-y}$.
- Multiply by e^y on both sides: $e^{2y} - xe^{2y} = 1 + x$. Note that $e^y e^y = e^{2y}$ and that $e^y e^{-y} = e^0 = 1$ because $x^m x^n = x^{m+n}$ (Chapter 1).
- Factor the e^{2y} out on the left: $e^{2y}(1 - x) = 1 + x$.

- Divide by $1-x$ on both sides: $e^{2y} = \frac{1+x}{1-x}$. (Since $-1 < x < 1$ for $\tanh^{-1} x$, we won't be in danger of dividing by zero.)
- Take the natural logarithm of both sides: $\ln(e^{2y}) = \ln\left(\frac{1+x}{1-x}\right)$.
- Apply the cancellation equation (Chapter 3): $2y = \ln\left(\frac{1+x}{1-x}\right)$.
- Divide by 2 on both sides: $y = \frac{1}{2}\ln\left(\frac{1+x}{1-x}\right)$
- Recall that $y = \tanh^{-1} x$ from the first step: $\tanh^{-1} x = \frac{1}{2}\ln\left(\frac{1+x}{1-x}\right)$.

3) $\operatorname{sech}^{-1} x = \ln\left(\frac{1}{x} + \sqrt{\frac{1}{x^2} - 1}\right)$

- Let $y = \operatorname{sech}^{-1} x$ such that $\operatorname{sech} y = x$.
- Since $\operatorname{sech} y = \frac{2}{e^y + e^{-y}}$, we may rewrite the previous equation as $\frac{2}{e^y + e^{-y}} = x$.
- Multiply by $e^y + e^{-y}$ on both sides: $2 = xe^y + xe^{-y}$.
- Subtract 2 from both sides: $0 = xe^y - 2 + xe^{-y}$.
- Multiply by e^y on both sides: $0 = xe^{2y} - 2e^y + x$. Note that $e^y e^y = e^{2y}$ and that $e^y e^{-y} = e^0 = 1$ because $x^m x^n = x^{m+n}$ (Chapter 1).
- $0 = xe^{2y} - 2e^y + x$ a quadratic equation in e^y. Apply the quadratic formula with $a = x$, $b = -2$, and $c = x$ to solve for e^y:

$$e^y = \frac{-b \pm \sqrt{b^2 - 4ac}}{2a} = \frac{-(-2) \pm \sqrt{(-2)^2 - 4(x)(x)}}{2(x)} = \frac{2 \pm \sqrt{4 - 4x^2}}{2x}$$

- Factor the 4 out: $\sqrt{4 - 4x^2} = \sqrt{4(1 - x^2)} = \sqrt{4}\sqrt{1 - x^2} = 2\sqrt{1 - x^2}$.
- Only the plus sign meets the domain requirement of $0 < x \leq 1$ for $\operatorname{sech}^{-1} x$.

$$e^y = \frac{2 + 2\sqrt{1 - x^2}}{2x} = \frac{2}{2x} + \frac{2\sqrt{1-x^2}}{2x} = \frac{1}{x} + \frac{\sqrt{1-x^2}}{x}$$

- Bring the x into the radical: $\frac{\sqrt{1-x^2}}{x} = \frac{\sqrt{1-x^2}}{\sqrt{x^2}} = \sqrt{\frac{1-x^2}{x^2}} = \sqrt{\frac{1}{x^2} - \frac{x^2}{x^2}} = \sqrt{\frac{1}{x^2} - 1}$.

$$e^y = \frac{1}{x} + \sqrt{\frac{1}{x^2} - 1}$$

- Take the natural logarithm of both sides: $\ln(e^y) = \ln\left(\frac{1}{x} + \sqrt{\frac{1}{x^2} - 1}\right)$.
- Apply the cancellation equation (Chapter 3): $y = \ln\left(\frac{1}{x} + \sqrt{\frac{1}{x^2} - 1}\right)$.
- Recall that $y = \operatorname{sech}^{-1} x$ from the first step: $\operatorname{sech}^{-1} x = \ln\left(\frac{1}{x} + \sqrt{\frac{1}{x^2} - 1}\right)$.

Chapter 7 Graphs

Exercise Set 7.1

1) $y = \frac{x^3}{8}$ for $-4 \leq x \leq 4$

x	-4	-2	0	2	4
y	-8	-1	0	1	8

x and y may each vary from $-\infty$ to ∞.

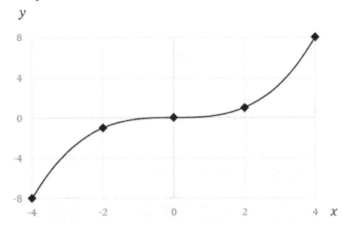

2) $y = \frac{x^4}{16}$ for $-4 \leq x \leq 4$

x	-4	-2	0	2	4
y	16	1	0	1	16

x may vary from $-\infty$ to ∞, while y ranges from 0 to ∞.

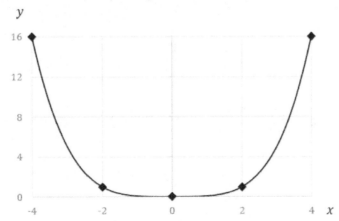

3) $y = x^{3/2}$ for $0 \leq x \leq 4$

We used a calculator to make the following table.

x	0	1	2	3	4
y	0	1	2.8	5.2	8

x and y may each vary from 0 to ∞; y isn't real for negative x.

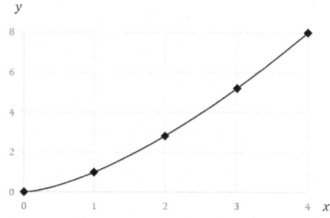

4) $y = x^{1/3}$ for $-64 \leq x \leq 64$

We used a calculator to make the following table.

x	−64	−32	0	32	64
y	−4	−3.2	0	3.2	4

x and y may each vary from $-\infty$ to ∞.

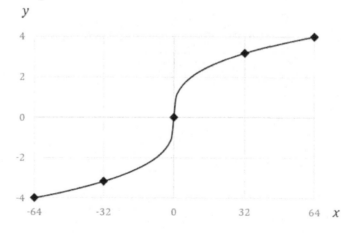

Logarithms and Exponentials Essential Skills Practice Workbook with Answers

Exercise Set 7.2

1) $y = \log_{10} x$ for $0 < x \leq 10$

We used a calculator to make the following table. As with all of the graphs in this book, we kept more digits than are shown in the table when making the actual graph.

x	0.2	0.4	0.6	0.8	1	3	5	7	10
y	−0.7	−0.4	−0.2	−0.1	0	0.5	0.7	0.8	1

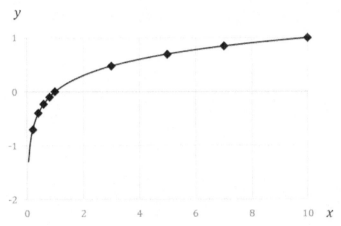

2) $y = \log_2 x$ for $0 < x \leq 8$

We used a calculator and the change of base formula to make the following table.

$$\log_2 x = \frac{\log_{10} x}{\log_{10} 2} \quad \text{or} \quad \log_2 x = \frac{\ln x}{\ln 2}$$

x	0.2	0.4	0.6	0.8	1	3	5	7	8
y	−2.3	−1.3	−0.7	−0.3	0	1.6	2.3	2.8	3

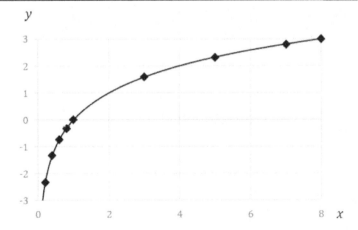

Exercise Set 7.3

1) $y = 1 - e^{-x}$ for $0 \leq x \leq 5$

We used a calculator to make the following table.

x	0	1	2	3	4	5
y	0	0.63	0.86	0.95	0.98	0.99

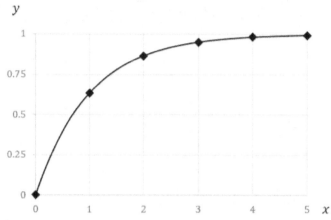

2) $y = 2^x$ for $-5 \leq x \leq 5$

We used a calculator to make the following table.

x	−5	−2.5	0	2.5	5
y	0.031	0.18	1	5.7	32

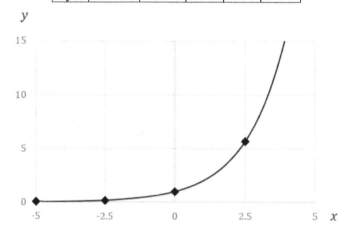

3) $y = e^{(x^2)}$ for $-2 \leq x \leq 2$

We used a calculator to make the following table.

x	−2	−1	0	1	2
y	55	2.7	1	2.7	55

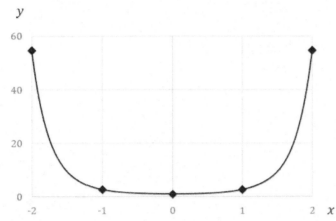

4) $y = e^{-(x^2)}$ for $-4 \leq x \leq 4$

We used a calculator to make the following table.

x	−4	−2	0	2	4
y	10^{-7}	0.018	1	0.018	10^{-7}

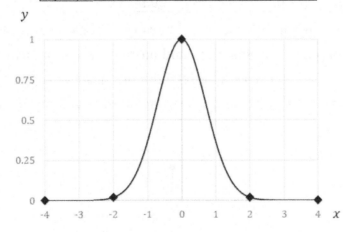

5) $y = \frac{1}{1+e^x}$ for $-5 \leq x \leq 5$

We used a calculator to make the following table.

x	−5	−2.5	0	2.5	5
y	0.99	0.92	0.5	0.076	0.007

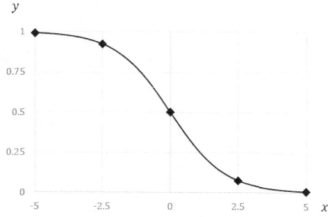

6) $y = \frac{1}{1-e^{-x}}$ for $-5 \leq x \leq 5$

We used a calculator to make the following table.

x	−5	−2.5	−0.1	0.1	2.5	5
y	−0.007	−0.089	−9.5	10.5	1.1	1

Note: This function isn't defined for $x = 0$ because this causes division by zero. As x approaches 0 from the left, y gets more and more negative (approaching $-\infty$). As x approaches 0 from the right, y gets more and more positive (approaching ∞). Thus, the limit doesn't exist at $x = 0$ (being $-\infty$ from one side and ∞ from the other).

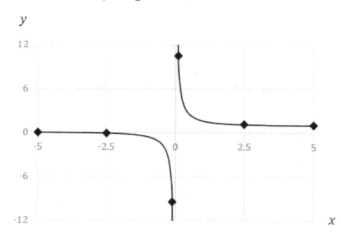

Exercise Set 7.4

1) $y = \text{sech } x$ for $-5 \leq x \leq 5$

We used a calculator to make the following table.

x	-5	-2.5	0	2.5	5
y	0.01	0.16	1	0.16	0.001

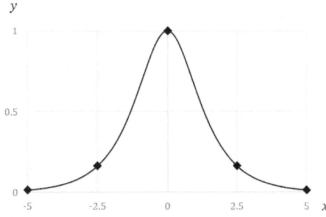

2) $y = \text{csch } x$ for $-5 \leq x \leq 5$

We used a calculator to make the following table.

x	-5	-2.5	-0.1	0.1	2.5	5
y	-0.01	-0.17	-10	10	0.17	0.01

Note: Since $\text{csch } x = \dfrac{2}{e^x - e^{-x}}$, this function isn't defined for $x = 0$ because this value of x results in division by zero.

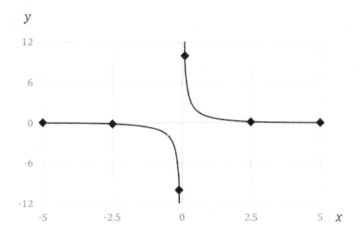

Answer Key

3) $y = \tanh x$ for $-2 \le x \le 2$

We used a calculator to make the following table.

Note: As with all of the graphs in this book, we used more data points than just the ones shown to draw the smooth curves. The representative points listed in the tables correspond to the diamond markers shown on the graphs.

x	-2	-1	0	1	2
y	-0.96	-0.76	0	0.76	0.96

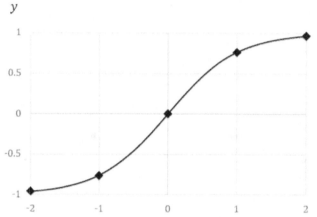

4) $y = \coth x$ for $-2 \le x \le 2$

We used a calculator to make the following table.

x	-2	-1	-0.1	0.1	1	2
y	-1	-1.3	-10	10	1.3	1

Notes: Since $\coth x = \dfrac{e^x + e^{-x}}{e^x - e^{-x}}$, this function isn't defined for $x = 0$ because this value of x results in division by zero; $y = \coth x$ approaches $y = \pm 1$ as x approaches $\pm\infty$.

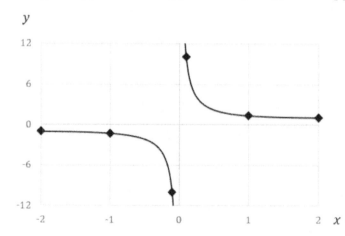

5) $y = \sinh^{-1} x$ for $-4 \leq x \leq 4$

We used a calculator to make the following table.

x	−4	−2	0	2	4
y	−2.1	−1.4	0	1.4	2.1

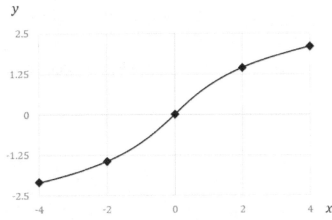

6) $y = \cosh^{-1} x$ for $0 < x \leq 5$

We used a calculator to make the following table.

x	1	2	3	4	5
y	0	1.3	1.8	2.1	2.3

Note: Since $\cosh x = \frac{e^x + e^{-x}}{2}$ can't be less than 1 (since this numerator can't be smaller than the denominator), the domain of $\cosh^{-1} x$ is $x \geq 1$. As x approaches 1, $\cosh^{-1} x$ approaches 0. (The **range** of $\cosh x$ corresponds to the **domain** of $\cosh^{-1} x$. See Sec. 6.7.)

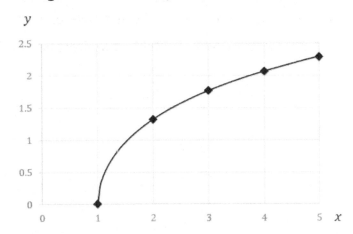

Answer Key

7) $y = \tanh^{-1} x$ for $-1 < x < 1$

We used a calculator to make the following table.

x	−0.99	−0.5	0	0.5	0.99
y	−2.6	−0.5	0	0.5	2.6

Note: Since $\tanh x = \frac{e^x - e^{-x}}{e^x + e^{-x}}$ can't exceed one (since this numerator can't be larger than the denominator), the domain of $\tanh^{-1} x$ is $-1 < x < 1$. As x approaches ± 1, $\tanh^{-1} x$ approaches $\pm \infty$. (The **range** of $\tanh x$ corresponds to the **domain** of $\tanh^{-1} x$.)

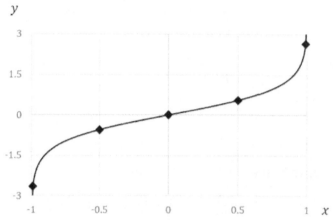

8) $y = \text{sech}^{-1} x$ for $0 < x \leq 1$

We used a calculator to make the following table.

x	0.01	0.2	0.4	0.6	0.8	1
y	5.3	2.3	1.6	1.1	0.7	0

Note: Since $\text{sech } x = \frac{2}{e^x + e^{-x}}$ can't exceed one or be negative (or even zero), the domain of $\text{sech}^{-1} x$ is $0 < x \leq 1$, As x approaches 0, $\text{sech}^{-1} x$ approaches $\pm \infty$.

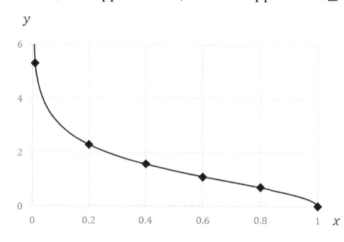

Logarithms and Exponentials Essential Skills Practice Workbook with Answers

Exercise Set 7.5

1) $y = e^{-x}$ for $0 \leq x \leq 8$

We used a calculator to make the following table.

x	0	2	4	6	8
y	1	0.14	0.018	0.0025	0.00034

Note: The logarithmic scaling linearized e^{-x} with negative slope since $\ln(e^{-x}) = -x$.

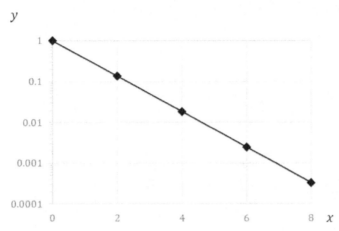

2) $y = 2^x$ for $0 \leq x \leq 10$

We made the following table.

x	0	2	4	6	8	10
y	1	4	16	64	256	1024

Note: As with the exponential curve, the logarithmic scaling linearizes 2^x.

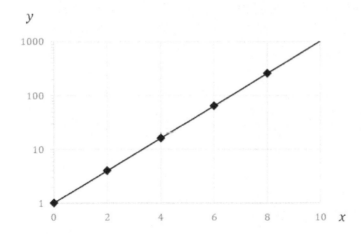

3) $y = \dfrac{1}{1-e^{-x}}$ for $0 < x \leq 1$

We used a calculator to make the following table.

x	0.01	0.2	0.4	0.6	0.8	1
y	101	5.5	3.0	2.2	1.8	1.6

Note: This branch of the graph still resembles the solution to Problem 6 of Sec. 7.3, even after logarithmic scaling. Note that y approaches ∞ as x approaches zero.

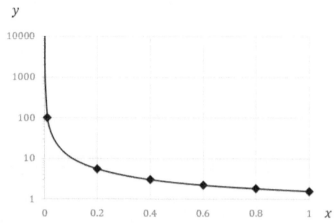

4) $y = x^2$ for $1 \leq x \leq 100$

We made the following table.

x	1	20	40	60	80	100
y	1	400	1600	3600	6400	10,000

Note: The logarithmic scaling inverted the concavity of the parabola, making a graph of x^2 look logarithmic. It effectively transformed x^2 into $\log_{10}(x^2) = 2\log_{10} x$.

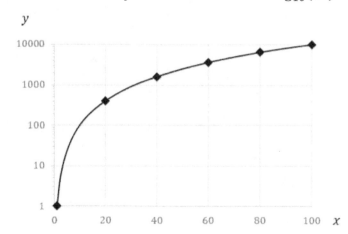

Chapter 8 Applications

Exercise Set 8.1

1) The initial population is: $N_0 = 720$ cells
(A) For part A only, the time is: $t = 0.25$ hr. (Note that 15 min. = 0.25 hr. Divide the minutes by 60 to convert from minutes to hours.)
When $t = 0.25$ hr, the population will be $N = 2(720) = 1440$ cells. This applies to Part A only.
Use the population formula: $N = N_0 e^{kt}$ becomes $1440 = 720 e^{k0.25}$
Divide by 720 on both sides: $2 = e^{0.25k}$
Take the natural logarithm of both sides: $\ln 2 = 0.25k$
Multiply by 4 on both sides: $4 \ln 2 = k \approx \boxed{2.77}$/hr.
Check: $N = N_0 e^{kt} \approx 720 e^{2.77(0.25)} \approx 720 e^{0.693} \approx 1440$
(B) In Part B only, the time is: $t = 4$ hr.
Use the population formula: $N = N_0 e^{kt} \approx 720 e^{2.77(4)} \approx 720 e^{11.08} \approx \boxed{4.67 \times 10^7}$ cells
(C) In Part C only, the final population is: $N = 1{,}000{,}000$ cells
Use the population formula: $N = N_0 e^{kt}$ becomes $1{,}000{,}000 \approx 720 e^{2.77 t}$
Divide by 720 on both sides: $1389 \approx e^{2.77k}$
Take the natural logarithm of both sides: $\ln 1389 \approx 2.77k$
Divide by 2.77 on both sides: $\frac{\ln 1389}{2.77} \approx t \approx \boxed{2.61}$ hr.
Check: $N = N_0 e^{kt} \approx 720 e^{2.77(2.61)} \approx 720 e^{7.23} \approx 1{,}000{,}000$

Note: We could model this problem with $N = 16^t (720)$ instead of $N = 720 e^{0.25t}$. With this model, we would get the same answers for parts B and C:
- $N = 16^4 (720) \approx 4.72 \times 10^7$ cells agrees with 4.67×10^7 cells (it would match more precisely if we kept additional digits throughout each calculation)
- $N = 16^{2.61}(720) \approx 1{,}000{,}000$ cells

The alternative formula, $N = 16^t (720)$, represents that the population doubles every 0.25 hours (which is 15 minutes). This means that the population multiplies by 16 every hour (since it doubles 4 times per hour), which is where the 16^t comes from.

Answer Key

Exercise Set 8.2

1) The initial mass is: $m_0 = 500$ g
(A) For Part A only, the time is $t = 80$ days
When $t = 80$ days, the mass of thorium-234 is $m = 50$ g. This applies to Part A only. (Convert 10% to 0.1 by dividing by 100%. Multiply 500 g by 0.1 to determine that the mass of the sample is 50 g after 80 days.)
Use the radioactive decay formula: $m = m_0 e^{-kt}$ becomes $50 = 500 e^{-k80}$
Divide by 500 on both sides: $0.1 = e^{-80k}$
Take the natural logarithm of both sides: $\ln 0.1 = -80k$
Divide by -80 on both sides: $\frac{\ln 0.1}{-80} = k \approx \boxed{0.0288}$/day
Note: $\ln 0.1 \approx -2.3$ such that $\frac{\ln 0.1}{-80}$ is positive.
Check: $m = m_0 e^{-kt} = 500 e^{-0.0288(80)} = 500 e^{-2.304} \approx 50$ g
(B) Use the formula that relates the decay constant to the half-life. Use the answer to Part B: $t_{½} = \frac{\ln 2}{k} \approx \frac{\ln 2}{0.0288} = \boxed{24.1}$ days
(C) The mass is now: $m = 2.5$ g
Use the radioactive decay formula: $m = m_0 e^{-kt}$ becomes $2.5 = 500 e^{-0.0288t}$
Divide by 500 on both sides: $0.005 = e^{-0.0288t}$
Take the natural logarithm of both sides: $\ln 0.005 = -0.0288t$
Divide by -0.0288 on both sides: $\frac{\ln 0.005}{-0.0288} = t \approx \boxed{184}$ days
Check: $m = m_0 e^{-kt} = 500 e^{-0.0288(184)} = 500 e^{-5.30} \approx 2.5$ g

Exercise Set 8.3

1) Given: $C = 6 \times 10^{-4}$ F, $Q_m = 3 \times 10^{-3}$ C, $R = 25$ Ω
(A) $\tau = RC = (25)(6 \times 10^{-4}) = \boxed{0.015}$ s
(B) $t_{½} = \tau \ln 2 = (0.015) \ln 2 \approx \boxed{0.010}$ s
(C) $Q = Q_m e^{-t/\tau} \to 1 \times 10^{-4} = (3 \times 10^{-3}) e^{-t/0.015} \to 0.0333 = e^{-t/0.015}$
$\to \ln(0.0333) = -\frac{t}{0.015} \to -3.4 = -\frac{t}{0.015} \to \boxed{0.051}$ s $= t$
Check: $Q = Q_m e^{-t/\tau} = (3 \times 10^{-3}) e^{-0.051/0.015} = (3 \times 10^{-3}) e^{-3.4} \approx 1 \times 10^{-4}$ C
(D) $Q = Q_m e^{-t/\tau} = (3 \times 10^{-3}) e^{-0.03/0.015} = (3 \times 10^{-3}) e^{-2} \approx \boxed{4.1 \times 10^{-4}}$ C

Logarithms and Exponentials Essential Skills Practice Workbook with Answers

2) Given: $C = 4 \times 10^{-4}$ F, $\Delta V_B = 10$ V, $R = 50$ Ω

(A) $Q_m = C\Delta V_B = (4 \times 10^{-4})(10) = \boxed{4 \times 10^{-3}}$ C

$I_m = \frac{\Delta V_B}{R} = \frac{10}{50} = \frac{1}{5} = \boxed{0.2}$ A

According to $Q = Q_m(1 - e^{-t/\tau})$, the charge on the capacitor is initially zero and grows toward its maximum value (Q_m) as time approaches infinity. According to $I = I_m e^{-t/\tau}$, when $t = 0$ the initial current is $I_0 = I_m e^0 = I_m$ (the maximum current). The charge grows whereas the current decays for a resistor and capacitor in series with a battery.

(B) $\tau = RC = (50)(4 \times 10^{-4}) = \boxed{0.02}$ s

(C) In one half-life: $t_{1/2} = \tau \ln 2 = (0.02) \ln 2 \approx \boxed{0.0139}$ s

Check: $Q = Q_m(1 - e^{-t/\tau}) = (4 \times 10^{-3})(1 - e^{-0.0139/0.02}) = (4 \times 10^{-3})(1 - e^{-0.695})$
$= (4 \times 10^{-3})(1 - 0.499) = (4 \times 10^{-3})(0.501) \approx 2 \times 10^{-3}$ C

(D) $Q = Q_m(1 - e^{-t/\tau}) = (4 \times 10^{-3})(1 - e^{-0.05/0.02}) = (4 \times 10^{-3})(1 - e^{-2.5})$
$\approx (4 \times 10^{-3})(1 - 0.0821) \approx (4 \times 10^{-3})(0.918) \approx \boxed{3.67 \times 10^{-3}}$ C

$I = I_m e^{-t/\tau} = 0.2 e^{-0.05/0.02} = 0.2 e^{-2.5} \approx \boxed{0.0164}$ A

Note: A science class (and even some math classes) would be picky about the number of significant figures used in the answers. The purpose of this answer key is to help you check your answers (not to show you how many of the digits are significant).

Exercise Set 8.4

1) $L = 10 \log_{10}\left(\frac{I}{I_0}\right) = 10 \log_{10}\left(\frac{1 \times 10^{-3}}{1 \times 10^{-12}}\right) = 10 \log_{10}[10^{-3-(-12)}]$
$= 10 \log_{10}(10^{-3+12}) = 10 \log_{10}(10^9) = 10(9) = \boxed{90}$ dB

2) $I = I_0 10^{L/10} = I_0 10^{50/10} = I_0 10^5 = (1 \times 10^{-12}) 10^5 = \boxed{1 \times 10^{-7}}$ W/m²

3) It is 100,000 times louder.

$\frac{I_2}{I_1} = \frac{I_0 10^{L_2/10}}{I_0 10^{L_1/10}} = \frac{10^{80/10}}{10^{30/10}} = \frac{10^8}{10^3} = 10^{8-3} = 10^5 = \boxed{100{,}000}$

Answer Key

Exercise Set 8.5

1) $P = \$4500, r = 20\% = \frac{20\%}{100\%} = 0.2, t = 3$ years

$A = Pe^{rt} = \$4500e^{0.2(3)} = \$4500e^{0.6} = \boxed{\$8199.53}$

2) $r = 2.5\% = \frac{2.5\%}{100\%} = 0.025$ Since the balance triples, we know that $A = 3P$

Substitute $A = 3P$ into $A = Pe^{rt}$ to get: $3P = Pe^{0.025t}$

Divide by P on both sides: $3 = e^{0.025t}$

Take the natural logarithm of both sides: $\ln 3 = 0.025t$

Divide by 0.025 on both sides: $\frac{\ln 3}{0.025} = t \approx 43.94$ years

Chapter 9 Calculus

Exercise Set 9.1

1) $\lim_{x \to 0^+} \ln x = -\infty$ As x gets less positive, $\ln x$ gets more negative.

2) $\lim_{x \to 0} e^x = 1$ since $e^0 = 1$

3) $\lim_{x \to \infty} e^{-x} = 0$ As x grows larger, $e^{-x} = \frac{1}{e^x}$ gets smaller. For example, when $x = 100$, $e^{-x} = e^{-100} \approx 3.72 \times 10^{-44}$ is extremely tiny.

4) $\lim_{x \to -\infty} 2^{-x} = \infty$ As x gets more negative, 2^{-x} grows larger. For example, when $x = -20$, $2^{-x} = 2^{-(-20)} = 2^{20} = 1{,}048{,}576$. Note that $-x$ is positive for negative values of x. For example, for $x = -20$, we get $-x = -(-20) = 20$.

5) $\lim_{x \to 0^+} \operatorname{csch} x = \infty$ As x gets less positive, $\operatorname{csch} x = \frac{1}{\sinh x} = \frac{2}{e^x - e^{-x}}$ grows larger. For example, for $x = 0.001$, $\operatorname{csch} x = \frac{2}{e^{0.001} - e^{-0.001}} \approx 1000$.

6) $\lim_{x \to 0^-} \operatorname{csch} x = -\infty$ As x gets less negative, $\operatorname{csch} x = \frac{1}{\sinh x} = \frac{2}{e^x - e^{-x}}$ grows more negative. For example, when $x = -0.001$, $\operatorname{csch} x = \frac{2}{e^{-0.001} - e^{-(-0.001)}} = \frac{2}{e^{-0.001} - e^{0.001}} \approx -1000$.

7) $\lim_{x \to 0} \operatorname{sech}^{-1} x = \infty$ As x gets less positive, $\operatorname{sech}^{-1} x$ grows larger. For example, when $x = 0.001$, $\operatorname{sech}^{-1} 0.001 \approx 7.6$. One way to see this is to recall the formula $\operatorname{sech}^{-1} x = \ln\left(\frac{1}{x} + \sqrt{\frac{1}{x^2} - 1}\right)$ from Sec. 6.7 (which is valid for $0 < x \leq 1$). For example, $\ln\left(\frac{1}{0.001} + \sqrt{\frac{1}{0.001^2} - 1}\right) \approx 7.6$. Another way to see it is to review Chapter 7.

8) $\lim_{x \to 1^+} \cosh^{-1} x = 0$ As x gets smaller approaching 1, $\cosh^{-1} x$ gets smaller. For example, when $x = 1.001$, $\cosh^{-1} 1.001 \approx 0.0447$. As with the previous solution, you can see this using the formula from Chapter 6 or the graph from Chapter 7.

9) $\lim_{x \to 0} \frac{\sinh x}{x} = \frac{\frac{d}{dx} \sinh x \big|_{x=0}}{\frac{d}{dx} x \big|_{x=0}} = \frac{\cosh x |_{x=0}}{1} = \frac{\cosh 0}{1} = \frac{1}{1} = 1$

We used l'Hôpital's rule because $\sinh x$ and x each approach zero as $x \to 0$, and since zero divided by zero is indeterminate. For $\frac{d}{dx} \sinh x$, see Sec.'s 9.2-9.3.

Answer Key

10) $\lim_{x\to\infty} \dfrac{\ln x}{x^2} = \dfrac{\frac{d}{dx}\ln x\big|_{x=\infty}}{\frac{d}{dx}x^2\big|_{x=\infty}} = \dfrac{\frac{1}{x}\big|_{x=\infty}}{2x\big|_{x=\infty}} = \dfrac{1}{x} \div 2x \bigg|_{x=\infty} = \dfrac{1}{2x^2}\bigg|_{x=\infty} = 0$

We used l'Hôpital's rule because $\ln x$ and x each approach infinity as $x \to \infty$, and since this is an indeterminate form. For $\dfrac{d}{dx}\ln x$, see Sec.'s 9.2-9.3.

Exercise Set 9.2

1) $\dfrac{e^{0.01}-1}{0.01} \approx \dfrac{1.01005-1}{0.01} \approx \dfrac{0.01005}{0.01} \approx 1.005$

2) $\dfrac{e^{0.001}-1}{0.001} \approx \dfrac{1.0010005-1}{0.001} \approx \dfrac{0.0010005}{0.001} \approx 1.0005$

3) $\dfrac{e^{0.0001}-1}{0.0001} \approx \dfrac{1.000100005-1}{0.0001} \approx \dfrac{0.000100005}{0.0001} \approx 1.00005$

4) $\dfrac{e^{0.00001}-1}{0.00001} \approx \dfrac{1.00001000005-1}{0.00001} \approx \dfrac{0.00001000005}{0.00001} \approx 1.000005$

Exercise Set 9.3

1) Recall from Chapter 6 that $\cosh kx = \dfrac{e^{kx}+e^{-kx}}{2}$ and $\sinh kx = \dfrac{e^{kx}-e^{-kx}}{2}$.

$\dfrac{d}{dx}\cosh kx = \dfrac{d}{dx}\dfrac{e^{kx}+e^{-kx}}{2} = \dfrac{1}{2}\dfrac{d}{dx}e^{kx} + \dfrac{1}{2}\dfrac{d}{dx}e^{-kx} = \dfrac{k}{2}e^{kx} - \dfrac{k}{2}e^{-kx} = k\dfrac{e^{kx}-e^{-kx}}{2} = \boxed{k\sinh kx}$

Note: The signs are different from the ordinary trig derivatives. Although $\dfrac{d}{dx}\cos x = -\sin x$, for hyperbolic functions we get $\dfrac{d}{dx}\cosh x = \sinh x$ without any minus sign.

2) Recall from Chapter 6 that $\tanh kx = \dfrac{\sinh kx}{\cosh kx} = \dfrac{e^{kx}-e^{-kx}}{e^{kx}+e^{-kx}}$ and $\operatorname{sech} kx = \dfrac{1}{\cosh kx} = \dfrac{2}{e^{kx}+e^{-kx}}$

$\dfrac{d}{dx}\tanh kx = \dfrac{d}{dx}\dfrac{\sinh kx}{\cosh kx} = \dfrac{\cosh kx \frac{d}{dx}\sinh kx - \sinh kx \frac{d}{dx}\cosh kx}{\cosh^2 kx} = \dfrac{k\cosh^2 kx - k\sinh^2 kx}{\cosh^2 kx} = \dfrac{k}{\cosh^2 kx} = \boxed{k\operatorname{sech}^2 k}$

We applied the quotient rule (Sec. 9.4). Recall from Chapter 6 that $\cosh^2 kx - \sinh^2 kx = 1$ (Sec. 6.2, Problem 8) and $\operatorname{sech} kx = \dfrac{1}{\cosh kx}$.

3) Recall from Chapter 6 that $\operatorname{sech} kx = \dfrac{1}{\cosh kx} = \dfrac{2}{e^{kx}+e^{-kx}}$ and $\tanh kx = \dfrac{\sinh kx}{\cosh kx} = \dfrac{e^{kx}-e^{-kx}}{e^{kx}+e^{-kx}}$

$\dfrac{d}{dx}\operatorname{sech} kx = \dfrac{d}{dx}\dfrac{1}{\cosh kx} = -\dfrac{1}{\cosh^2 kx}\dfrac{d}{dx}\cosh kx = -\dfrac{k\sinh kx}{\cosh^2 kx} = \boxed{-k\operatorname{sech} kx \tanh kx}$

We applied the chain rule (Sec. 9.4), $\dfrac{df}{dx} = \dfrac{df}{du}\dfrac{du}{dx}$, with $u = \cosh kx$ and $f = \dfrac{1}{u}$, to write

$\dfrac{d}{dx}\dfrac{1}{\cosh kx} = \dfrac{df}{dx} = \dfrac{df}{du}\dfrac{du}{dx} = \dfrac{d}{du}\dfrac{1}{u}\dfrac{d}{dx}\cosh kx = -\dfrac{1}{u^2}k\sinh kx = -\dfrac{k\sinh kx}{\cosh^2 kx}$. We applied the rule $\dfrac{d}{dx}x^n = nx^{n-1}$ to get $\dfrac{d}{du}\dfrac{1}{u} = \dfrac{d}{du}u^{-1} = (-1)u^{-2} = -\dfrac{1}{u^2}$.

See the note regarding signs from the solution to Problem 1. The signs for hyperbolic functions are a little different from the signs of the ordinary trig functions.

4) Recall from Chapter 6 that $\operatorname{csch} kx = \frac{1}{\sinh kx} = \frac{2}{e^{kx}-e^{-kx}}$ and $\coth kx = \frac{\cosh kx}{\sinh kx} = \frac{e^{kx}+e^{-kx}}{e^{kx}-e^{-kx}}$.

$\frac{d}{dx}\operatorname{csch} kx = \frac{d}{dx}\frac{1}{\sinh kx} = -\frac{1}{\sinh^2 kx}\frac{d}{dx}\sinh kx = -\frac{k\cosh kx}{\sinh^2 kx} = \boxed{-k\operatorname{csch} kx \coth kx}$

We applied the chain rule (Sec. 9.4), $\frac{df}{dx} = \frac{df}{du}\frac{du}{dx}$, with $u = \sinh kx$ and $f = \frac{1}{u}$, to write

$\frac{d}{dx}\frac{1}{\sinh kx} = \frac{df}{dx} = \frac{df}{du}\frac{du}{dx} = \frac{d}{du}\frac{1}{u}\frac{d}{dx}\sinh kx = -\frac{1}{u^2}k\cosh kx = -\frac{k\cosh kx}{\sinh^2 kx}$. We applied the

rule $\frac{d}{dx}x^n = nx^{n-1}$ to get $\frac{d}{du}\frac{1}{u} = \frac{d}{du}u^{-1} = (-1)u^{-2} = -\frac{1}{u^2}$.

5) Recall from Chapter 6 that $\coth kx = \frac{\cosh kx}{\sinh kx} = \frac{e^{kx}+e^{-kx}}{e^{kx}-e^{-kx}}$ and $\operatorname{csch} kx = \frac{1}{\sinh kx} = \frac{2}{e^{kx}-e^{-kx}}$.

$\frac{d}{dx}\coth kx = \frac{d}{dx}\frac{\cosh kx}{\sinh kx} = \frac{\sinh kx \frac{d}{dx}\cosh kx - \cosh kx \frac{d}{dx}\sinh kx}{\sinh^2 kx} = \frac{k\sinh^2 kx - k\cosh^2 kx}{\sinh^2 kx} = \frac{-k}{\cosh^2 kx} = \boxed{-k\operatorname{csch}^2 kx}$

We applied the quotient rule (Sec. 9.4). Recall from Chapter 6 that $\cosh^2 kx - \sinh^2 kx = 1$ (Sec. 6.2, Problem 8), such that $\sinh^2 kx - \cosh^2 kx = -1$, and $\operatorname{csch} kx = \frac{1}{\sinh kx}$.

6) Use the cancellation equations from Chapter 3 to write $b = e^{\ln b}$ and the property from Chapter 1 that $(y^m)^n = y^{mn}$. Also, recall that $\frac{d}{dx}e^{kx} = ke^{kx}$; let $k = \ln b$.

$\frac{d}{dx}b^x = \frac{d}{dx}(e^{\ln b})^x = \frac{d}{dx}e^{x\ln b} = (\ln b)e^{x\ln b} = (\ln b)e^{\ln b^x} = \boxed{(\ln b)b^x}$

7) Use the change of base formula from Chapter 4: $\log_{10} x = \frac{\ln x}{\ln 10}$. Also, recall that $\frac{d}{dx}\ln x = \frac{1}{x}$.

$\frac{d}{dx}\log_{10} x = \frac{d}{dx}\frac{\ln x}{\ln 10} = \frac{1}{\ln 10}\frac{d}{dx}\ln x = \boxed{\frac{1}{x\ln 10}}$

8) Use either method from Example 4. We will use the alternate solution.
Let $y = \cosh^{-1} x$ such that $\cosh y = x$. Take a derivative of both sides:

$$\frac{d}{dx}\cosh y = \frac{d}{dx}x$$

Use the chain rule, $\frac{df}{dx} = \frac{df}{du}\frac{du}{dx}$ (Sec. 9.4), with $f = \cosh y$ and $u = y$ such that $\frac{df}{dx} = \frac{df}{du}\frac{du}{dx}$ is $\frac{d}{dx}\cosh y = \frac{d}{dy}\cosh y \frac{dy}{dx}$. Note that $\frac{d}{dy}\cosh y = \sinh y$ and $\frac{d}{dx}x = 1$.

$$\sinh y \frac{dy}{dx} = 1$$

Divide both sides by $\sinh y$.

$$\frac{dy}{dx} = \frac{1}{\sinh y}$$

Answer Key

Recall from Sec. 6.2, Problem 8, that $\cosh^2 y - \sinh^2 y = 1$, such that $\cosh^2 y = 1 + \sinh^2 y$ and $\cosh^2 y - 1 = \sinh^2 y$. Square root both sides: $\sqrt{\cosh^2 y - 1} = \sinh y$.

$$\frac{dy}{dx} = \frac{1}{\sqrt{\cosh^2 y - 1}}$$

Recall that $y = \cosh^{-1} x$ and $\cosh y = x$.

$$\frac{d}{dx}\cosh^{-1} x = \frac{1}{\sqrt{x^2 - 1}}$$

9) Use either method from Example 4. We will use the alternate solution.

Let $y = \tanh^{-1} x$ such that $\tanh y = x$. Take a derivative of both sides:

$$\frac{d}{dx}\tanh y = \frac{d}{dx}x$$

Use the chain rule, $\frac{df}{dx} = \frac{df}{du}\frac{du}{dx}$ (Sec. 9.4), with $f = \tanh y$ and $u = y$ such that $\frac{df}{dx} = \frac{df}{du}\frac{du}{dx}$ is $\frac{d}{dx}\tanh y = \frac{d}{dy}\tanh y \frac{dy}{dx}$. Note that $\frac{d}{dy}\tanh y = \operatorname{sech}^2 y$ (Problem 2) and $\frac{d}{dx}x = 1$.

$$\operatorname{sech}^2 y \frac{dy}{dx} = 1$$

Divide both sides by $\operatorname{sech}^2 y$.

$$\frac{dy}{dx} = \frac{1}{\operatorname{sech}^2 y}$$

Recall from Sec. 6.4, Example 1, that $1 - \tanh^2 x = \operatorname{sech}^2 x$.

$$\frac{dy}{dx} = \frac{1}{1 - \tanh^2 x}$$

Recall that $y = \tanh^{-1} x$ and $\tanh y = x$.

$$\frac{d}{dx}\tanh^{-1} x = \frac{1}{1 - x^2}$$

10) Use either method from Example 4. We will use the alternate solution.

Let $y = \operatorname{sech}^{-1} x$ such that $\operatorname{sech} y = x$. Take a derivative of both sides:

$$\frac{d}{dx}\operatorname{sech} y = \frac{d}{dx}x$$

Use the chain rule, $\frac{df}{dx} = \frac{df}{du}\frac{du}{dx}$ (Sec. 9.4), with $f = \operatorname{sech} y$ and $u = y$ such that $\frac{df}{dx} = \frac{df}{du}\frac{du}{dx}$ is $\frac{d}{dx}\operatorname{sech} y = \frac{d}{dy}\operatorname{sech} y \frac{dy}{dx}$. Note that $\frac{d}{dy}\operatorname{sech} y = -\operatorname{sech} y \tanh y$ (Problem 3) and $\frac{d}{dx}x = 1$.

$$-\operatorname{sech} y \tanh y \frac{dy}{dx} = 1$$

Divide both sides by $-\text{sech}\, y \tanh y$.

$$\frac{dy}{dx} = -\frac{1}{\text{sech}\, y \tanh y}$$

Recall from Sec. 6.4, Example 1, that $1 - \tanh^2 y = \text{sech}^2 y$, such that $1 = \text{sech}^2 y + \tanh^2 y$ and $1 - \text{sech}^2 y = \tanh^2 y$. Square root both sides: $\sqrt{1 - \text{sech}^2 y} = \tanh y$.

$$\frac{dy}{dx} = -\frac{1}{\text{sech}\, y \sqrt{1 - \text{sech}^2 y}}$$

Recall that $y = \text{sech}^{-1} x$ and $\text{sech}\, y = x$.

$$\frac{d}{dx} \text{sech}^{-1} x = -\frac{1}{x\sqrt{1 - x^2}}$$

Exercise Set 9.4

1) Use the chain rule with $u = x^2$ and $f = e^u$. Recall from Sec. 9.3 that $\frac{d}{du} e^u = e^u$.

$$\frac{d}{dx} e^{(x^2)} = \frac{df}{du}\frac{du}{dx} = \frac{d}{du} e^u \frac{d}{dx} x^2 = e^u (2x) = \boxed{2x e^{(x^2)}}$$

2) Use the product rule with $f = x$ and $g = \ln x$. Recall from Sec. 9.3 that $\frac{d}{dx} \ln x = \frac{1}{x}$.

$$\frac{d}{dx} x \ln x = \ln x \frac{d}{dx} x + x \frac{d}{dx} \ln x = (\ln x)(1) + x\left(\frac{1}{x}\right) = \ln x + 1 = \boxed{1 + \ln x}$$

3) Use the quotient rule with $f = 1 + \ln x$ and $g = 1 - \ln x$. Recall from Sec. 9.3 that $\frac{d}{dx} \ln x = \frac{1}{x}$.

$$\frac{d}{dx} \frac{1+\ln x}{1-\ln x} = \frac{(1-\ln x)\frac{d}{dx}(1+\ln x) - (1+\ln x)\frac{d}{dx}(1-\ln x)}{(1-\ln x)^2} = \frac{(1-\ln x)\left(\frac{1}{x}\right) - (1+\ln x)\left(-\frac{1}{x}\right)}{(1-\ln x)^2} \text{ Note: } -1\left(-\frac{1}{x}\right) = \frac{1}{x}$$

$$= \frac{(1-\ln x)\left(\frac{1}{x}\right) + (1+\ln x)\left(\frac{1}{x}\right)}{(1-\ln x)^2} = \frac{(1-\ln x + 1 + \ln x)\left(\frac{1}{x}\right)}{(1-\ln x)^2} = \frac{1-\ln x + 1 + \ln x}{x(1-\ln x)^2} = \boxed{\frac{2}{x(1-\ln x)^2}}$$

4) Use the chain rule with $u = kx$ and $f = \ln u$. Recall from Sec. 9.3 that $\frac{d}{du} \ln u = \frac{1}{u}$.

$$\frac{d}{dx} \ln(kx) = \frac{df}{du}\frac{du}{dx} = \frac{d}{du} \ln u \frac{d}{dx} kx = \frac{1}{u}(k) = \frac{k}{kx} = \boxed{\frac{1}{x}}$$

Note: The constant k cancels out.

Alternate solution: $\frac{d}{dx} \ln(kx) = \frac{d}{dx}(\ln k + \ln x) = \frac{d}{dx} \ln k + \frac{d}{dx} \ln x = 0 + \frac{1}{x} = \frac{1}{x}$

In the alternate solution, we used the formula $\ln(kx) = \ln k + \ln x$ from Sec. 3.2.

Answer Key

5) Use the chain rule with $u = \text{sech}\, x$ and $f = u^2$. Recall from Sec. 9.3 that $\frac{d}{dx}\text{sech}\, x = -\text{sech}\, x \tanh x$ (see Problem 3).

$\frac{d}{dx}\text{sech}^2 x = \frac{df}{du}\frac{du}{dx} = \frac{d}{du}u^2 \frac{d}{dx}\text{sech}\, x = 2u(-\text{sech}\, x \tanh x) = \boxed{-2\,\text{sech}^2 x \tanh x}$

Alternate answer: $\boxed{-\frac{2\sinh x}{\cosh^3 x}}$

6) Use the product rule with $f = \sinh x$ and $g = \cosh x$. Recall from Sec. 9.3 that $\frac{d}{dx}\sinh x = \cosh x$ and $\frac{d}{dx}\cosh x = \sinh x$. (As discussed in the solutions to Sec. 9.3, both of these derivatives are positive, whereas for the ordinary trig functions one of these similar derivatives is negative.)

$\frac{d}{dx}\sinh x \cosh x = \cosh x \frac{d}{dx}\sinh x + \sinh x \frac{d}{dx}\cosh x = \boxed{\cosh^2 x + \sinh^2 x}$

Note: This **doesn't** simplify to 1. Although for ordinary trig functions, $\sin^2 x + \cos^2 x$ equals one, the similar hyperbolic identity is different: $\cosh^2 x - \sinh^2 x = 1$ (Sec. 6.2, Problem 8).

7) Use the chain rule with $u = \cos x$ and $f = \tanh^{-1} u$. Recall from trigonometry that $\frac{d}{dx}\cos x = -\sin x$ (this is an ordinary cosine; it isn't a hyperbolic function) and from Sec. 9.3 that $\frac{d}{du}\tanh^{-1} u = \frac{1}{1-u^2}$ (see Problem 9). Also recall from trigonometry that $\sin^2 x + \cos^2 x = 1$, such that $\sin^2 x = 1 - \cos^2 x$.

$\frac{d}{dx}\tanh^{-1}(\cos x) = \frac{df}{du}\frac{du}{dx} = \frac{d}{du}\tanh^{-1} u \frac{d}{dx}\cos x = \frac{1}{1-u^2}(-\sin x)$

$= \frac{-\sin x}{1-\cos^2 x} = \frac{-\sin x}{\sin^2 x} = \boxed{-\frac{1}{\sin x}} = \boxed{-\csc x}$

8) Use the chain rule with $u = \tanh x$ and $f = \sqrt{u}$. Recall from calculus that $\frac{d}{du}\sqrt{u} = \frac{d}{du}u^{1/2} = \frac{1}{2}u^{-1/2} = \frac{1}{2u^{1/2}} = \frac{1}{2\sqrt{u}}$ and from Sec. 9.3 that $\frac{d}{dx}\tanh x = \text{sech}^2 x$ (see Problem 2).

$\frac{d}{dx}\sqrt{\tanh x} = \frac{df}{du}\frac{du}{dx} = \frac{d}{du}\sqrt{u}\frac{d}{dx}\tanh x = \frac{1}{2\sqrt{u}}\text{sech}^2 x = \boxed{\frac{\text{sech}^2 x}{2\sqrt{\tanh x}}}$

9) Use the chain rule with $u = e^x$ and $f = \sinh u$. Recall from Sec. 9.3 that $\frac{d}{du}\sinh u = \cosh u$ and that $\frac{d}{dx}e^x = e^x$.

$\frac{d}{dx}\sinh(e^x) = \frac{df}{du}\frac{du}{dx} = \frac{d}{du}\sinh u \frac{d}{dx}e^x = (\cosh u)(e^x) = \boxed{e^x \cosh(e^x)}$

Logarithms and Exponentials Essential Skills Practice Workbook with Answers

10) Use the cancellation equation to write $x = e^{\ln x}$. Then use the rule $(y^m)^n = y^{mn}$ from Sec. 1.5 to write $(e^{\ln x})^x = e^{x \ln x}$. Then use the chain rule with $u = x \ln x$ and $f = e^u$. Recall from Sec. 9.3 that $\frac{d}{du} e^u = e^u$. Finally, use the product rule with x and $\ln x$. Recall from Sec. 9.3 that $\frac{d}{dx} \ln x = \frac{1}{x}$.

$\frac{d}{dx} x^x = \frac{d}{dx} (e^{\ln x})^x = \frac{d}{dx} e^{x \ln x} = \frac{df}{du} \frac{du}{dx} = \frac{d}{du} e^u \frac{d}{dx} (x \ln x)$

$= e^u \left(\ln x \frac{d}{dx} x + x \frac{d}{dx} \ln x \right) = e^{x \ln x} \left[(\ln x)(1) + x \left(\frac{1}{x} \right) \right]$

$= e^{x \ln x} (\ln x + 1) = e^{\ln(x^x)} (\ln x + 1) = x^x (\ln x + 1) = \boxed{x^x (1 + \ln x)}$

Note: In the beginning of the last line, we used the rule $y \ln x = \ln(x^y)$ from Sec. 3.4 to write $e^{x \ln x} = e^{\ln(x^x)}$ and then we used the cancellation equation again to write $e^{\ln(x^x)} = x^x$.

Exercise Set 9.5

1) $e^{-1} \approx 1 + (-1) + \frac{(-1)^2}{2} + \frac{(-1)^3}{6} + \frac{(-1)^4}{24} + \frac{(-1)^5}{120} + \frac{(-1)^6}{720} + \frac{(-1)^7}{5040}$

$= 1 - 1 + \frac{1}{2} - \frac{1}{6} + \frac{1}{24} - \frac{1}{120} + \frac{1}{720} - \frac{1}{5040} \approx 0.367857143$

Compare to the actual value: $e^{-1} = \frac{1}{e} \approx 0.367879441$

2) $e^2 \approx 1 + 2 + \frac{2^2}{2} + \frac{2^3}{6} + \frac{2^4}{24} + \frac{2^5}{120} + \frac{2^6}{720} + \frac{2^7}{5040}$

$= 1 + 2 + \frac{4}{2} + \frac{8}{6} + \frac{16}{24} + \frac{32}{120} + \frac{64}{720} + \frac{128}{5040} \approx 7.380952381$

Compare to the actual value: $e^2 \approx 7.389056099$

3) $\sin\left(\frac{\pi}{6}\right) \approx x - \frac{x^3}{6} + \frac{x^5}{120} = \frac{\pi}{6} - \frac{\pi^3}{(216)(6)} + \frac{\pi^5}{(7776)(120)}$

$\approx 0.523598776 - 0.023924596 + 0.000327953 \approx 0.500002133$

Compare to the actual value: $\sin\left(\frac{\pi}{6}\right) = \sin 30° = \frac{1}{2} = 0.5$

4) $\ln 2 = \ln(1 + 1) \approx 1 - \frac{1^2}{2} + \frac{1^3}{3} - \frac{1^4}{4} + \frac{1^5}{5} - \frac{1^6}{6} + \frac{1^7}{7}$

$= 1 - \frac{1}{2} + \frac{1}{3} - \frac{1}{4} + \frac{1}{5} - \frac{1}{6} + \frac{1}{7} = \frac{319}{420} = 0.75952381$

Compare to the actual value: $\ln 2 \approx 0.693147181$
(This case doesn't converge as rapidly.)

Answer Key

Exercise Set 9.6

1) Recall the solution to Problem 6 of Sec. 9.3: $\frac{d}{dx} b^x = (\ln b) b^x$.

It follows that $\frac{d}{dx} \frac{b^x}{\ln b} = \frac{1}{\ln b} \frac{d}{dx} b^x = b^x$, which means that $\int 2^x \, dx = \boxed{\frac{2^x}{\ln 2} + c}$.

Check: $\frac{d}{dx}\left(\frac{2^x}{\ln 2} + c\right) = \frac{d}{dx} \frac{2^x}{\ln 2} + \frac{d}{dx} c = \frac{1}{\ln 2} \frac{d}{dx} 2^x + 0 = \frac{1}{\ln 2}(\ln 2) 2^x = 2^x$

2) Recall from Chapter 6 that $\coth x = \frac{\cosh x}{\sinh x}$.

Let $u = \sinh x$ such that $du = \cosh x \, dx$ (since $\frac{d}{dx} \sinh x = \cosh x$).

$\int \coth x \, dx = \int \frac{\cosh x}{\sinh x} dx = \int \frac{du}{u} = \ln u + c = \boxed{\ln|\sinh x| + c}$

Check: $\frac{d}{dx}(\ln|\sinh x| + c) = \frac{1}{\sinh x} \frac{d}{dx} \sinh x = \frac{\cosh x}{\sinh x} = \coth x$

3) The integral $\int \text{sech } x \, dx$ is fascinating in the regard that you can obtain answers that look considerably different, yet are mathematically equivalent, using different approaches. We'll solve this problem two different ways. At the end, the "check" will show that both solutions are equivalent.

Solution 1: Use the substitution $u = e^x$.

Recall from Chapter 6 that $\text{sech } x = \frac{1}{\cosh x} = \frac{2}{e^x + e^{-x}}$.

Let $u = e^x$ such that $du = e^x dx$ (since $\frac{d}{dx} e^x = e^x$). Multiply by $\frac{e^x}{e^x}$. Note that $e^x e^x = e^{2x} = u^2$ and $e^x e^{-x} = e^0 = 1$.

$\int \text{sech } x \, dx = \int \frac{2}{e^x + e^{-x}} dx = \int \frac{2}{e^x + e^{-x}} \frac{e^x}{e^x} dx = \int \frac{2 e^x dx}{e^{2x} + 1} = \int \frac{2 du}{u^2 + 1}$

Let $u = \tan \theta$ (ordinary, not hyperbolic) such that $du = \sec^2 \theta \, d\theta$.

Recall the (ordinary) trig identity $\tan^2 \theta + 1 = \sec^2 \theta$.

$\int \text{sech } x \, dx = \int \frac{2 \sec^2 \theta \, d\theta}{\tan^2 \theta + 1} = \int \frac{2 \sec^2 \theta \, d\theta}{\sec^2 \theta} = 2 \int d\theta = 2\theta + c$

Recall that $u = \tan \theta$ and $u = e^x$.

$\int \text{sech } x \, dx = 2 \tan^{-1} u + c = \boxed{2 \tan^{-1}(e^x) + c}$ (one possible answer; see the next page for an alternative)

Note that this is an ordinary inverse tangent; it's not hyperbolic. See the alternate solution that follows to see that this is equivalent to $\tan^{-1}(\sinh x) + c$.

Solution 2: Use the substitution $u = \sinh x$.

Recall from Chapter 6 that $\operatorname{sech} x = \frac{1}{\cosh x}$ and that $\cosh^2 x - \sinh^2 x = 1$ (see Problem 8 in Sec. 6.2; also note that the signs are different for this hyperbolic identity compared to the similar trig identity). It follows that $\cosh x = \sqrt{1 + \sinh^2 x}$.

Let $u = \sinh x$ such that $du = \cosh x\, dx$ (since $\frac{d}{dx} \sinh x = \cosh x$) and $dx = \frac{du}{\cosh x}$. Since $\cosh x = \sqrt{1 + \sinh^2 x}$, we may write $\cosh x = \sqrt{1 + u^2}$ and $dx = \frac{du}{\sqrt{1+u^2}}$.

$\int \operatorname{sech} x\, dx = \int \frac{1}{\cosh x} dx = \int \frac{1}{\sqrt{1+u^2}} \frac{du}{\sqrt{1+u^2}} = \int \frac{du}{1+u^2}$

Let $u = \tan \theta$ (ordinary, not hyperbolic) such that $du = \sec^2 \theta\, d\theta$.

Recall the (ordinary) trig identity $\tan^2 \theta + 1 = \sec^2 \theta$.

$\int \operatorname{sech} x\, dx = \int \frac{\sec^2 \theta\, d\theta}{1 + \tan^2 \theta} = \int \frac{\sec^2 \theta\, d\theta}{\sec^2 \theta} = \int d\theta = \theta + c$

Recall that $u = \tan \theta$ and $u = \sinh x$.

$\int \operatorname{sech} x\, dx = \theta + c = \tan^{-1} u + c = \boxed{\tan^{-1}|\sinh x| + c}$ (another possible answer)

(The absolute values reflect that $u > 0$ for $\ln u$.)

Recall from calculus that the derivative of the inverse tangent (also called arctangent) is $\frac{d}{dx} \tan^{-1} x = \frac{1}{1+x^2}$. Note that this is an ordinary inverse tangent; it's not hyperbolic.

Check for Solution 1: $\frac{d}{dx}[2\tan^{-1}(e^x) + c] = \frac{2}{1+(e^x)^2} \frac{d}{dx} e^x = \frac{2e^x}{1+e^{2x}} = \frac{2e^x}{1+e^{2x}} \frac{e^{-x}}{e^{-x}} = \frac{2}{e^{-x}+e^x} = \operatorname{sech} x$

Check for Solution 2: $\frac{d}{dx}(\tan^{-1}|\sinh x| + c) = \frac{1}{1+\sinh^2 x} \frac{d}{dx} \sinh x = \frac{\cosh x}{\cosh^2 x} = \frac{1}{\cosh x} = \operatorname{sech} x$

We used the hyperbolic identity $\cosh^2 x - \sinh^2 x = 1$ (see Problem 8 in Sec. 6.2) which is equivalent to $\cosh^2 x = 1 + \sinh^2 x$. In both checks, we applied the chain rule to take a derivative of the inverse tangent function. For example, for $\tan^{-1}|\sinh x|$, we treated it as $f = \tan^{-1} u$ with $u = \sinh x$ to write $\frac{df}{dx} = \frac{df}{du} \frac{du}{dx}$. In the check for Solution 1, we multiplied by $\frac{e^{-x}}{e^{-x}}$; note that $e^x e^{-x} = e^0 = 1$ and $e^{2x} e^{-x} = e^x$. Recall from Chapter 6 that $\operatorname{sech} x = \frac{1}{\cosh x} = \frac{2}{e^x + e^{-x}}$.

4) Recall from Chapter 6 that $\operatorname{csch} x = \frac{1}{\sinh x}$.

As with the solution to Problem 3, there is more than one way to perform $\int \operatorname{csch} x\, dx$, and the final answer may appear to look considerably different, yet be mathematically

Answer Key

equivalent. If you get a different answer and wish to see if your answer is equivalent, take a derivative of your answer and see if you can make it agree with csch x (just like we will do in the "check" that follows our solution).

For variety, we will solve this problem with a much different substitution than we used in the solutions to Problem 3. We will take advantage of a hyperbolic identity from Chapter 6.

Recall from Chapter 6 (Sec. 6.2, Problem 6) that $\sinh(2y) = 2 \sinh y \cosh y$. Let $y = \frac{x}{2}$ to see that $\sinh x = 2 \sinh\left(\frac{x}{2}\right) \cosh\left(\frac{x}{2}\right)$. Substitute this into $\operatorname{csch} x = \frac{1}{\sinh x}$ to get $\operatorname{csch} x = \frac{1}{2 \sinh\left(\frac{x}{2}\right) \cosh\left(\frac{x}{2}\right)}$.

$\int \operatorname{csch} x \, dx = \int \frac{dx}{\sinh x} = \int \frac{dx}{2 \sinh\left(\frac{x}{2}\right) \cosh\left(\frac{x}{2}\right)}$

Now divide the numerator and denominator each by $\cosh^2\left(\frac{x}{2}\right)$. The numerator will become $\operatorname{sech}^2\left(\frac{x}{2}\right)$ since $\frac{1}{\cosh^2\left(\frac{x}{2}\right)} = \operatorname{sech}^2\left(\frac{x}{2}\right)$ and the denominator will become $2 \tanh\left(\frac{x}{2}\right)$ since $\frac{2 \sinh\left(\frac{x}{2}\right) \cosh\left(\frac{x}{2}\right)}{\cosh^2\left(\frac{x}{2}\right)} = \frac{2 \sinh\left(\frac{x}{2}\right)}{\cosh\left(\frac{x}{2}\right)} = 2 \tanh\left(\frac{x}{2}\right)$.

$\int \operatorname{csch} x \, dx = \int \frac{\operatorname{sech}^2\left(\frac{x}{2}\right) dx}{2 \tanh\left(\frac{x}{2}\right)}$

Let $u = \tanh\left(\frac{x}{2}\right)$ such that $du = \frac{1}{2} \operatorname{sech}^2\left(\frac{x}{2}\right) dx$ since $\frac{d}{dx} \tanh kx = k \operatorname{sech}^2 kx$ (Sec. 9.3, Problem 2, with k equal to $\frac{1}{2}$). Note that $2du = \operatorname{sech}^2\left(\frac{x}{2}\right) dx$.

$\int \operatorname{csch} x \, dx = \int \frac{2 du}{2u} = \int \frac{du}{u} = \ln u + c$

Recall that $u = \tanh\left(\frac{x}{2}\right)$.

$\int \operatorname{csch} x \, dx = \boxed{\ln\left|\tanh\left(\frac{x}{2}\right)\right| + c}$ (The absolute values reflect that $u > 0$ for $\ln u$.)

Check: $\frac{d}{dx}\left(\ln\left|\tanh\left(\frac{x}{2}\right)\right| + c\right) = \frac{1}{\tanh\left(\frac{x}{2}\right)} \frac{d}{dx} \tanh\left(\frac{x}{2}\right) + \frac{d}{dx} c = \frac{1}{\tanh\left(\frac{x}{2}\right)} \operatorname{sech}^2\left(\frac{x}{2}\right) \frac{d}{dx}\left(\frac{x}{2}\right)$

$= \left[\frac{1}{\tanh\left(\frac{x}{2}\right)} \operatorname{sech}^2\left(\frac{x}{2}\right)\right]\left(\frac{1}{2}\right) = \frac{\cosh\left(\frac{x}{2}\right)}{2 \sinh\left(\frac{x}{2}\right) \cosh^2\left(\frac{x}{2}\right)} = \frac{1}{2 \sinh\left(\frac{x}{2}\right) \cosh\left(\frac{x}{2}\right)} = \frac{1}{\sinh x} = \operatorname{csch} x$

We used $\sinh x = 2 \sinh\left(\frac{x}{2}\right) \cosh\left(\frac{x}{2}\right)$ just like we did at the beginning of our solution.

5) Integrate by parts, $\int u\, dv = uv - \int v\, du$, with $u = x$ and $dv = \sinh x\, dx$ such that $du = dx$ and $v = \cosh x$ (since $\frac{d}{dx}\cosh x = \sinh x$).

$\int x \sinh x\, dx = x \cosh x - \int \cosh x\, dx = \boxed{x \cosh x - \sinh x + c}$

Note that the signs for the hyperbolic derivatives and integrals aren't exactly the same as the similar derivatives and integrals with ordinary trig functions (see the solution to Problem 1 in Sec. 9.3).

Check: $\frac{d}{dx}(x \cosh x - \sinh x + c) = \frac{d}{dx}(x \cosh x) - \frac{d}{dx}\sinh x + \frac{d}{dx}c$

$= \cosh x \frac{d}{dx}x + x \frac{d}{dx}\cosh x - \frac{d}{dx}\sinh x$

$= (\cosh x)(1) + x \sinh x - \cosh x = \cosh x + x \sinh x - \cosh x = x \sinh x$

We applied the product rule (Sec. 9.4) with $f = x$ and $g = \cosh x$.

6) Integrate by parts, $\int u\, dv = uv - \int v\, du$, with $u = x$ and $dv = e^x dx$ such that $du = dx$ and $v = e^x$ (since $\frac{d}{dx}e^x = e^x$).

$\int xe^x\, dx = xe^x - \int e^x\, dx = xe^x - e^x + c = \boxed{e^x(x - 1) + c}$

Check: $\frac{d}{dx}[e^x(x - 1) + c] = (x - 1)\frac{d}{dx}e^x + e^x \frac{d}{dx}(x - 1) + \frac{d}{dx}c$

$= (x - 1)e^x + e^x(1) = xe^x - e^x + e^x = xe^x$

We applied the product rule (Sec. 9.4) with $f = e^x$ and $g = x - 1$. Recall from calculus that $\frac{d}{dx}(x - 1) = \frac{d}{dx}x - \frac{d}{dx}1 = 1 - 0 = 1$.

7) Integrate by parts, $\int u\, dv = uv - \int v\, du$, with $u = \ln x$ and $dv = \frac{1}{x^2}$ such that $du = \frac{1}{x}dx$ (since $\frac{d}{dx}\ln x = \frac{1}{x}$) and $v = -\frac{1}{x}$ (since $\frac{d}{dx}\frac{1}{x} = \frac{d}{dx}x^{-1} = -x^{-2} = -\frac{1}{x^2}$).

$\int \frac{\ln x}{x^2} dx = (\ln x)\left(-\frac{1}{x}\right) - \int \left(-\frac{1}{x}\right)\frac{1}{x} dx = -\frac{\ln x}{x} + \int \frac{dx}{x^2}$

$= -\frac{\ln x}{x} - \frac{1}{x} + c = -\frac{1}{x}(\ln x + 1) + c = \boxed{-\frac{1}{x}(1 + \ln x) + c}$

Recall from calculus that $\int \frac{dx}{x^2} = -\frac{1}{x} + c$ because $\frac{d}{dx}\frac{1}{x} = \frac{d}{dx}x^{-1} = -\frac{1}{x^2}$.

Check: $\frac{d}{dx}\left[-\frac{1}{x}(1 + \ln x) + c\right] = -(1 + \ln x)\frac{d}{dx}\frac{1}{x} - \frac{1}{x}\frac{d}{dx}(1 + \ln x) + \frac{d}{dx}c$

$= -(1 + \ln x)\left(-\frac{1}{x^2}\right) - \frac{1}{x}\left(0 + \frac{1}{x}\right) = (1 + \ln x)\left(\frac{1}{x^2}\right) - \frac{1}{x}\left(\frac{1}{x}\right) = \frac{1}{x^2} + \frac{\ln x}{x^2} - \frac{1}{x^2} = \frac{\ln x}{x^2}$

We applied the product rule (Sec. 9.4) with $f = -\frac{1}{x}$ and $g = 1 + \ln x$.

Answer Key

Chapter 10 Complex Numbers

Exercise Set 10.1

1) $(9 + 6i)(7 - 8i) = 9(7) + 9(-8i) + 6i(7) + 6i(-8i)$
$= 63 - 72i + 42i - 48i^2 = 63 - 30i - 48(-1) = 63 + 48 - 30i = 111 - 30i$

2) $(7 - 9i)^2 = (7 - 9i)(7 - 9i) = 7(7) + 7(-9i) - 9i(7) - 9i(-9i)$
$= 49 - 63i - 63i + 81i^2 = 49 - 126i + 81(-1) = 49 - 81 - 126i = -32 - 126i$

3) $(1 - i)^3 = (1 - i)(1 - i)(1 - i) = (1 - i)[1(1) + 1(-i) - i(1) - i(-i)]$
$= (1 - i)(1 - i - i + i^2) = (1 - i)(1 - 2i - 1) = (1 - i)(-2i) = 1(-2i) - i(-2i)$
$= -2i + 2i^2 = -2i + 2(-1) = -2i - 2 = -2 - 2i$

4) $(4 + 9i)(4 + 9i)^* = (4 + 9i)(4 - 9i) = 4^2 + 9^2 = 16 + 81 = 97$
Use the formula $(x + iy)(x - iy) = x^2 + y^2$

5) $(7 - 6i)(7 - 6i)^* = (7 - 6i)(7 + 6i) = 7^2 + (-6)^2 = 49 + 36 = 85$

6) $\frac{2+i}{4+5i} = \frac{2+i}{4+5i} \frac{4-5i}{4-5i} = \frac{2(4)+2(-5i)+i(4)+i(-5i)}{4^2+5^2} = \frac{8-10i+4i-5i^2}{16+25} = \frac{8-6i-5(-1)}{41} = \frac{8-6i+5}{41} = \frac{13-6i}{41}$

7) $\frac{2-7i}{8-3i} = \frac{2-7i}{8-3i} \frac{8+3i}{8+3i} = \frac{2(8)+2(3i)-7i(8)-7i(3i)}{8^2+(-3)^2} = \frac{16+6i-56i-21i^2}{64+9} = \frac{16-50i-21(-1)}{73} = \frac{16-50i+21}{73} = \frac{37-50i}{73}$

8) $i^{14} = i^{12}i^2 = (i^4)^3 i^2 = 1^3(-1) = 1(-1) = \boxed{-1}$ Notes: $i^4 = 1$ and $i^2 = -1$

The main idea is to find the largest power smaller than 14 that is divisible by 4. This lets us take advantage of the fact that $i^4 = 1$. Recall from Sec. 1.5 that $(x^m)^n = x^{mn}$ and $x^m x^n = x^{m+n}$.

9) $i^{25} = i^{24} i^1 = (i^4)^6 i^1 = 1^6 i = 1i = \boxed{i}$ Notes: $i^4 = 1$ and $i^1 = 1$

10) $i^{2027} = i^{2024} i^3 = (i^4)^{506} i^3 = 1^{506}(-i) = 1(-i) = \boxed{-i}$ Notes: $i^4 = 1$ and $i^3 = -i$

11) $i^{1000} = (i^4)^{250} = 1^{250} = \boxed{1}$ Note: $i^4 = 1$

12) $\frac{1}{i} = \frac{1}{i} \frac{i}{i} = \frac{i}{i^2} = \frac{i}{-1} = \boxed{-i}$

13) $i^{-11} = i^{-12} i^1 = (i^4)^{-3} i^1 = 1^{-3} i = 1i = \boxed{i}$ Note: $i^4 = 1$

Exercise Set 10.2

1) $2 - 5i$

2 to the right, 5 down

$|z| = \sqrt{2^2 + (-5)^2} = \sqrt{4 + 25} = \sqrt{29}$

units from the origin

2) $4i$

0 to the right, 4 up

$|z| = \sqrt{0^2 + 4^2} = \sqrt{16} = 4$

units from the origin

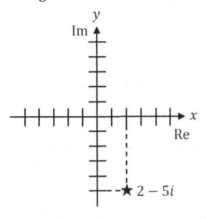
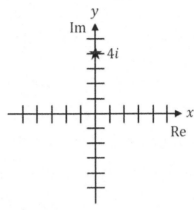

3) $-4 - 4i$

4 left, 4 down

$|z| = \sqrt{(-4)^2 + (-4)^2} = \sqrt{16 + 16}$
$= \sqrt{32} = \sqrt{16(2)} = \sqrt{16}\sqrt{2} = 4\sqrt{2}$

units from the origin

4) $-3 + 2i$

3 to the left, 2 up

$|z| = \sqrt{(-3)^2 + 2^2} = \sqrt{9 + 4} = \sqrt{13}$

units from the origin

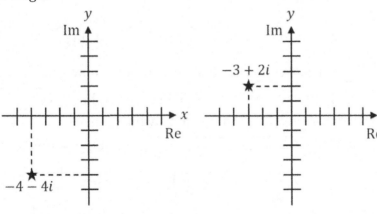

Answer Key

Exercise Set 10.3

1) $x = 6\sqrt{2}$ and $y = -6\sqrt{2}$

$$r = |z| = \sqrt{x^2 + y^2} = \sqrt{\left(6\sqrt{2}\right)^2 + \left(-6\sqrt{2}\right)^2}$$

$$= \sqrt{6^2\left(\sqrt{2}\right)^2 + 6^2\left(\sqrt{2}\right)^2} = \sqrt{36(2) + 36(2)} = \sqrt{72 + 72} = \sqrt{144} = 12$$

$$\theta = \tan^{-1}\left(\frac{y}{x}\right) = \tan^{-1}\left(\frac{-6\sqrt{2}}{6\sqrt{2}}\right) = \tan^{-1}(-1) = \frac{7\pi}{4} \text{ rad}$$

$$z = r(\cos\theta + i\sin\theta) = 12\left[\cos\left(\frac{7\pi}{4}\right) + i\sin\left(\frac{7\pi}{4}\right)\right]$$

Note: Unlike Example 1, here $x = 6\sqrt{2}$ is positive. Since $x = 6\sqrt{2}$ is positive and $y = -6\sqrt{2}$ is negative, the angle lies in Quadrant IV. The angle $\frac{7\pi}{4}$ is equivalent to $-\frac{\pi}{4}$, since it differs from $\frac{7\pi}{4}$ by 2π. An alternate answer is thus $z = 12\left[\cos\left(-\frac{\pi}{4}\right) + i\sin\left(-\frac{\pi}{4}\right)\right]$.

2) $x = -\frac{1}{4}$ and $y = -\frac{\sqrt{3}}{4}$

$$r = |z| = \sqrt{x^2 + y^2} = \sqrt{\left(-\frac{1}{4}\right)^2 + \left(-\frac{\sqrt{3}}{4}\right)^2} = \sqrt{\frac{1}{16} + \frac{3}{16}} = \sqrt{\frac{4}{16}} = \sqrt{\frac{1}{4}} = \frac{1}{2}$$

$$\theta = \tan^{-1}\left(\frac{y}{x}\right) = \tan^{-1}\left(\frac{-\frac{\sqrt{3}}{4}}{-\frac{1}{4}}\right) = \tan^{-1}\left(\frac{\sqrt{3}}{4} \div \frac{1}{4}\right) = \tan^{-1}\left(\frac{\sqrt{3}}{4} \times \frac{4}{1}\right) = \tan^{-1}(\sqrt{3}) = \frac{4\pi}{3} \text{ rad}$$

$$z = r(\cos\theta + i\sin\theta) = \frac{1}{2}\left[\cos\left(\frac{4\pi}{3}\right) + i\sin\left(\frac{4\pi}{3}\right)\right]$$

Note: Since $x = -\frac{1}{4}$ is negative, add π to the calculator's answer of $\frac{\pi}{3}$ to get $\frac{4\pi}{3}$ rad. See the note in Example 1. Since x and y are both negative, θ lies in Quadrant III.

3) $x = 0$ and $y = -5$ (the given number is purely imaginary)

$$r = |z| = \sqrt{x^2 + y^2} = \sqrt{0^2 + (-5)^2} = \sqrt{0 + 25} = \sqrt{25} = 5$$

$$\theta = \tan^{-1}\left(\frac{y}{x}\right) = \tan^{-1}\left(\frac{-5}{0}\right) = \frac{3\pi}{2} \text{ rad or } -\frac{\pi}{2} \text{ rad}$$

$$z = r(\cos\theta + i\sin\theta) = 5\left[\cos\left(\frac{3\pi}{2}\right) + i\sin\left(\frac{3\pi}{2}\right)\right]$$

Note: Although $\frac{y}{x} = -\frac{5}{0}$ is undefined, the angle is a definite $\frac{3\pi}{2}$ rad. We can determine this geometrically because $x = 0$ and $y = -5$, corresponding to the $-y$-axis (at 270°).

4) $x = -9$ and $y = 0$ (the given number is purely real)
$$r = |z| = \sqrt{x^2 + y^2} = \sqrt{(-9)^2 + 0^2} = \sqrt{81 + 0} = \sqrt{81} = 9$$
$$\theta = \tan^{-1}\left(\frac{y}{x}\right) = \tan^{-1}\left(\frac{0}{-9}\right) = \tan^{-1}(0) = \pi \text{ rad}$$
$$z = r(\cos\theta + i\sin\theta) = 9[\cos(\pi) + i\sin(\pi)]$$

Note: θ lies along the $-x$-axis, which is π counterclockwise from $+x$. Since $x = -9$ is negative, add π to the calculator's answer of 0 to get π rad. See the note in Example 1.

Exercise Set 10.4

1) $e^{i\pi/3} = \cos\left(\frac{\pi}{3}\right) + i\sin\left(\frac{\pi}{3}\right) = \frac{1}{2} + \frac{\sqrt{3}}{2}i$

2) $e^{3i\pi/4} = \cos\left(\frac{3\pi}{4}\right) + i\sin\left(\frac{3\pi}{4}\right) = -\frac{\sqrt{2}}{2} + \frac{\sqrt{2}}{2}i$ Note that $\frac{\sqrt{2}}{2} = \frac{1}{\sqrt{2}}$

3) $e^{5i\pi/6} = \cos\left(\frac{5\pi}{6}\right) + i\sin\left(\frac{5\pi}{6}\right) = -\frac{\sqrt{3}}{2} + \frac{i}{2}$

4) $e^{-1+3i\pi/2} = e^{-1}\left[\cos\left(\frac{3\pi}{2}\right) + i\sin\left(\frac{3\pi}{2}\right)\right] = \frac{1}{e}[0 + i(-1)] = -\frac{i}{e}$

5) $e^{2+i\pi} = e^2[\cos\pi + i\sin\pi] = e^2(-1 + 0i) = -e^2$

6) $e^{1-4i\pi/3} = e^1\left[\cos\left(-\frac{4\pi}{3}\right) + i\sin\left(-\frac{4\pi}{3}\right)\right] = e\left(-\frac{1}{2} + \frac{\sqrt{3}}{2}i\right) = -\frac{e}{2} + \frac{e\sqrt{3}}{2}i$

7) $x = \sqrt{3}$ and $y = -1$
$$r = |z| = \sqrt{x^2 + y^2} = \sqrt{\left(\sqrt{3}\right)^2 + (-1)^2} = \sqrt{3+1} = \sqrt{4} = 2$$
$$\theta = \tan^{-1}\left(\frac{y}{x}\right) = \tan^{-1}\left(\frac{-1}{\sqrt{3}}\right) = -\frac{\pi}{6} \text{ rad or } \frac{11\pi}{6} \text{ rad}$$
$$z = re^{i\theta} = 2e^{-i\pi/6} \text{ or } 2e^{11i\pi/6}$$

Note: Since $x = \sqrt{3}$ is positive, a calculator will give the correct angle for the inverse tangent. Since x is positive and y is negative, the angle lies in Quadrant IV. The angle $-\frac{\pi}{6}$ is equivalent to $\frac{11\pi}{6}$, since they differ by 2π. An alternate answer is thus $z = 2e^{11i\pi/6}$.

8) $x = 0$ and $y = -7$ (the given number is purely imaginary)
$$r = |z| = \sqrt{x^2 + y^2} = \sqrt{0^2 + (-7)^2} = \sqrt{0 + 49} = \sqrt{49} = 7$$
$$\theta = \tan^{-1}\left(\frac{y}{x}\right) = \tan^{-1}\left(\frac{-7}{0}\right) = \frac{3\pi}{2} \text{ rad or } -\frac{\pi}{2} \text{ rad}$$
$$z = re^{i\theta} = 7e^{3i\pi/2} \text{ or } 7e^{-i\pi/2}$$

Answer Key

Note: Although $\frac{y}{x} = -\frac{7}{0}$ is undefined, the angle is a definite $\frac{3\pi}{2}$ rad. We can determine this geometrically because $x = 0$ and $y = -7$, corresponding to the $-y$-axis (at 270°).

Exercise Set 10.5

1) $x = 3$ and $y = 3$ Since x and y are both positive, θ lies in Quadrant I.

$r = |z| = \sqrt{x^2 + y^2} = \sqrt{3^2 + 3^2} = \sqrt{9 + 9} = \sqrt{18} = \sqrt{(9)(2)} = \sqrt{9}\sqrt{2} = 3\sqrt{2}$

$\theta = \tan^{-1}\left(\frac{y}{x}\right) = \tan^{-1}\left(\frac{3}{3}\right) = \frac{\pi}{4}$ rad

$z^n = r^n[\cos(n\theta) + i\sin(n\theta)] = (3 + 3i)^6 = \left(3\sqrt{2}\right)^6\left[\cos\left(\frac{6\pi}{4}\right) + i\sin\left(\frac{6\pi}{4}\right)\right]$

$(3 + 3i)^6 = 3^6\left(\sqrt{2}\right)^6\left[\cos\left(\frac{3\pi}{2}\right) + i\sin\left(\frac{3\pi}{2}\right)\right] = 729(8)(0 - i) = \boxed{-5832i}$

Since $n = 6$, note that $n\theta = 6\left(\frac{\pi}{4}\right) = \frac{6\pi}{4} = \frac{3\pi}{2}$.

2) $x = \sqrt{3}$ and $y = 1$ Since x and y are both positive, θ lies in Quadrant I.

$r = |z| = \sqrt{x^2 + y^2} = \sqrt{\left(\sqrt{3}\right)^2 + 1^2} = \sqrt{3 + 1} = \sqrt{4} = 2$

$\theta = \tan^{-1}\left(\frac{y}{x}\right) = \tan^{-1}\left(\frac{1}{\sqrt{3}}\right) = \frac{\pi}{6}$ rad

$z^n = r^n[\cos(n\theta) + i\sin(n\theta)] = \left(\sqrt{3} + i\right)^{12} = 2^{12}\left[\cos\left(\frac{12\pi}{6}\right) + i\sin\left(\frac{12\pi}{6}\right)\right]$

$(3 + 3i)^{12} = 4096[\cos(2\pi) + i\sin(2\pi)] = 4096(1 + 0i) = \boxed{4096}$

Since $n = 12$, note that $n\theta = 12\left(\frac{\pi}{6}\right) = \frac{12\pi}{6} = 2\pi$.

3) $x = -\frac{1}{2}$ and $y = \frac{\sqrt{3}}{2}$ Since x is negative while y is positive, θ lies in Quadrant II.

$r = |z| = \sqrt{x^2 + y^2} = \sqrt{\left(-\frac{1}{2}\right)^2 + \left(\frac{\sqrt{3}}{2}\right)^2} = \sqrt{\frac{1}{4} + \frac{3}{4}} = \sqrt{1} = 1$

$\theta = \tan^{-1}\left(\frac{y}{x}\right) = \tan^{-1}\left(\frac{\sqrt{3}/2}{-1/2}\right) = \tan^{-1}\left(-\frac{\sqrt{3}}{2} \div \frac{1}{2}\right) = \tan^{-1}\left(-\frac{\sqrt{3}}{2} \times \frac{2}{1}\right) = \tan^{-1}(-\sqrt{3}) = \frac{2\pi}{3}$ rad

$z^n = r^n[\cos(n\theta) + i\sin(n\theta)] = \left(-\frac{1}{2} + i\frac{\sqrt{3}}{2}\right)^3 = 1^3\left[\cos\left(\frac{6\pi}{3}\right) + i\sin\left(\frac{6\pi}{3}\right)\right]$

$\left(-\frac{1}{2} + i\frac{\sqrt{3}}{2}\right)^3 = 1[\cos(2\pi) + i\sin(2\pi)] = 1(1 + 0i) = 1(1) = \boxed{1}$

Note: Since x is negative, a calculator would give $-\frac{\pi}{3}$ rad. Add π to get $-\frac{\pi}{3} + \pi = \frac{2\pi}{3}$ to put the angle in Quadrant II. Since $n = 3$, note that $n\theta = 3\left(\frac{2\pi}{3}\right) = \frac{6\pi}{3} = 2\pi$.

4) $x = \sqrt{2}$ and $y = -\sqrt{2}$ Since x is positive while y is negative, θ lies in Quadrant IV.

$r = |z| = \sqrt{x^2 + y^2} = \sqrt{(\sqrt{2})^2 + (-\sqrt{2})^2} = \sqrt{2+2} = \sqrt{4} = 2$

$\theta = \tan^{-1}\left(\frac{y}{x}\right) = \tan^{-1}\left(\frac{-\sqrt{2}}{\sqrt{2}}\right) = \tan^{-1}(-1) = -\frac{\pi}{4}$ rad or $\frac{7\pi}{4}$ rad

$z^n = r^n[\cos(n\theta) + i\sin(n\theta)] = (\sqrt{2} - i\sqrt{2})^7 = 2^7\left[\cos\left(-\frac{7\pi}{4}\right) + i\sin\left(-\frac{7\pi}{4}\right)\right]$

$(\sqrt{2} - i\sqrt{2})^7 = 128\left(\frac{\sqrt{2}}{2} + \frac{\sqrt{2}}{2}i\right) = \frac{128\sqrt{2}}{2}(1+i) = \boxed{64\sqrt{2} + 64i\sqrt{2}}$

Note that $-\frac{\pi}{4}$ rad is equivalent to $\frac{7\pi}{4}$ rad since they differ by 2π. Since $n = 7$, note that $n\theta = 7\left(-\frac{\pi}{4}\right) = -\frac{7\pi}{4}$. Also note that $-\frac{7\pi}{4}$ is equivalent to $\frac{\pi}{4}$ since they differ by 2π.

Exercise Set 10.6

1) $x = -1$ and $y = -\sqrt{3}$ Since x and y are both negative, θ lies in Quadrant III.

$r = |z| = \sqrt{x^2 + y^2} = \sqrt{(-1)^2 + (-\sqrt{3})^2} = \sqrt{1+3} = \sqrt{4} = 2$

$\theta = \tan^{-1}\left(\frac{y}{x}\right) = \tan^{-1}\left(\frac{-\sqrt{3}}{-1}\right) = \tan^{-1}(\sqrt{3}) = \frac{4\pi}{3}$ rad

Note: Since x is negative, a calculator would give $\frac{\pi}{3}$ rad. Add π to get $\frac{\pi}{3} + \pi = \frac{4\pi}{3}$ to put the angle in Quadrant III. For the fourth roots, $n = 4$.

$u = r^{1/n}\left[\cos\left(\frac{\theta + 2\pi k}{n}\right) + i\sin\left(\frac{\theta + 2\pi k}{n}\right)\right] = 2^{1/4}\left[\cos\left(\frac{4\pi/3 + 2\pi k}{4}\right) + i\sin\left(\frac{4\pi/3 + 2\pi k}{4}\right)\right]$

$u = 2^{1/4}\left[\cos\left(\frac{\pi}{3} + \frac{2\pi k}{4}\right) + i\sin\left(\frac{\pi}{3} + \frac{2\pi k}{4}\right)\right] = 2^{1/4}\cos\left(\frac{\pi}{3} + \frac{2\pi k}{4}\right) + 2^{1/4}i\sin\left(\frac{\pi}{3} + \frac{2\pi k}{4}\right)$

For $k = 0$: $\quad u = 2^{1/4}\cos\left(\frac{\pi}{3}\right) + i2^{1/4}\sin\left(\frac{\pi}{3}\right) = \frac{2^{1/4}}{2} + \frac{2^{1/4}\sqrt{3}}{2}i = \frac{1}{2^{3/4}} + \frac{\sqrt{3}}{2^{3/4}}i$

For $k = 1$: $\quad u = 2^{1/4}\cos\left(\frac{\pi}{3} + \frac{\pi}{2}\right) + i2^{1/4}\sin\left(\frac{\pi}{3} + \frac{\pi}{2}\right) = \cos\left(\frac{5\pi}{6}\right) + i2^{1/4}\sin\left(\frac{5\pi}{6}\right) =$

$-\frac{2^{1/4}\sqrt{3}}{2} + \frac{2^{1/4}}{2}i = -\frac{\sqrt{3}}{2^{3/4}} + \frac{i}{2^{3/4}}$

For $k = 2$: $\quad u = 2^{1/4}\cos\left(\frac{\pi}{3} + \pi\right) + i2^{1/4}\sin\left(\frac{\pi}{3} + \pi\right) = \cos\left(\frac{4\pi}{3}\right) + i2^{1/4}\sin\left(\frac{4\pi}{3}\right) =$

$-\frac{2^{1/4}}{2} - \frac{2^{1/4}\sqrt{3}}{2}i = -\frac{1}{2^{3/4}} - \frac{\sqrt{3}}{2^{3/4}}i$

For $k = 3$: $\quad u = 2^{1/4}\cos\left(\frac{\pi}{3} + \frac{3\pi}{2}\right) + i2^{1/4}\sin\left(\frac{\pi}{3} + \frac{3\pi}{2}\right) = \cos\left(\frac{11\pi}{6}\right) + i2^{1/4}\sin\left(\frac{11\pi}{6}\right) =$

$\frac{2^{1/4}\sqrt{3}}{2} - \frac{2^{1/4}}{2}i = \frac{\sqrt{3}}{2^{3/4}} - \frac{i}{2^{3/4}}$

Answer Key

You can check these answers by multiplying them out, though keeping track of all of the exponents and signs can get a little tricky. For example:

$$\left(\frac{1}{2^{3/4}} + \frac{\sqrt{3}}{2^{3/4}}i\right)^4 = \left(\frac{1}{2^{3/4}} + \frac{\sqrt{3}}{2^{3/4}}i\right)\left(\frac{1}{2^{3/4}} + \frac{\sqrt{3}}{2^{3/4}}i\right)\left(\frac{1}{2^{3/4}} + \frac{\sqrt{3}}{2^{3/4}}i\right)\left(\frac{1}{2^{3/4}} + \frac{\sqrt{3}}{2^{3/4}}i\right)$$

$$= \left(\frac{1}{2^{3/2}} + \frac{2\sqrt{3}}{2^{3/2}}i - \frac{3}{2^{3/2}}\right)\left(\frac{1}{2^{3/2}} + \frac{2\sqrt{3}}{2^{3/2}}i - \frac{3}{2^{3/2}}\right) = \left(-\frac{2}{2^{3/2}} + \frac{i\sqrt{3}}{2^{1/2}}\right)\left(-\frac{2}{2^{3/2}} + \frac{i\sqrt{3}}{2^{1/2}}\right)$$

$$= \left(-\frac{1}{2^{1/2}} + \frac{i\sqrt{3}}{2^{1/2}}\right)\left(-\frac{1}{2^{1/2}} + \frac{i\sqrt{3}}{2^{1/2}}\right) = \frac{1}{2^1} - \frac{2i\sqrt{3}}{2^1} - \frac{3}{2^1} = -\frac{2}{2} - i\sqrt{3} = -1 - i\sqrt{3}$$

2) $x = 0$ and $y = 1$ Since x is zero while y is positive, $\theta = \frac{\pi}{2}$ rad.

$r = |z| = \sqrt{x^2 + y^2} = \sqrt{0^2 + 1^2} = \sqrt{0 + 1} = \sqrt{1} = 1$

$\theta = \tan^{-1}\left(\frac{y}{x}\right) = \tan^{-1}\left(\frac{1}{0}\right) = \frac{\pi}{2}$ rad

Note: Although $\frac{y}{x} = \frac{1}{0}$ is undefined, the angle is a definite $\frac{\pi}{2}$ rad. We can determine this geometrically because $x = 0$ and $y = 1$, corresponding to the $+y$-axis (at 90°).

For the square roots, $n = 2$.

$u = r^{1/n}\left[\cos\left(\frac{\theta + 2\pi k}{n}\right) + i\sin\left(\frac{\theta + 2\pi k}{n}\right)\right] = 1^{1/2}\left[\cos\left(\frac{\pi/2 + 2\pi k}{2}\right) + i\sin\left(\frac{\pi/2 + 2\pi k}{2}\right)\right]$

$u = 1\left[\cos\left(\frac{\pi}{4} + \pi k\right) + i\sin\left(\frac{\pi}{4} + \pi k\right)\right] = \cos\left(\frac{\pi}{4} + \pi k\right) + i\sin\left(\frac{\pi}{4} + \pi k\right)$

For $k = 0$: $u = \cos\left(\frac{\pi}{4}\right) + i\sin\left(\frac{\pi}{4}\right) = \frac{\sqrt{2}}{2} + \frac{\sqrt{2}}{2}i$ Alternate answers: $\pm\left(\frac{1}{\sqrt{2}} + \frac{i}{\sqrt{2}}\right)$

For $k = 1$: $u = \cos\left(\frac{\pi}{4} + \pi\right) + i\sin\left(\frac{\pi}{4} + \pi\right) = \cos\left(\frac{5\pi}{4}\right) + i\sin\left(\frac{5\pi}{4}\right) = -\frac{\sqrt{2}}{2} - \frac{\sqrt{2}}{2}i$

Check these answers by squaring them. For example:

$$\left(\frac{\sqrt{2}}{2} + \frac{\sqrt{2}}{2}i\right)^2 = \left(\frac{\sqrt{2}}{2} + \frac{\sqrt{2}}{2}i\right)\left(\frac{\sqrt{2}}{2} + \frac{\sqrt{2}}{2}i\right) = \frac{2}{4} + 2\left(\frac{2}{4}i\right) - \frac{2}{4} = 0 + \frac{4}{4}i = i$$

3) $x = 27$ and $y = 0$ Since x is positive while y is zero, $\theta = 0$ rad.

$r = |z| = \sqrt{x^2 + y^2} = \sqrt{27^2 + 0^2} = \sqrt{27^2 + 0} = \sqrt{27^2} = 27$

$\theta = \tan^{-1}\left(\frac{y}{x}\right) = \tan^{-1}\left(\frac{0}{27}\right) = \tan^{-1}(0) = 0$ rad For the cube roots, $n = 3$.

$u = r^{1/n}\left[\cos\left(\frac{\theta + 2\pi k}{n}\right) + i\sin\left(\frac{\theta + 2\pi k}{n}\right)\right] = 27^{1/3}\left[\cos\left(\frac{0 + 2\pi k}{3}\right) + i\sin\left(\frac{0 + 2\pi k}{3}\right)\right]$

$u = 3\left[\cos\left(\frac{2\pi k}{3}\right) + i\sin\left(\frac{2\pi k}{3}\right)\right] = 3\cos\left(\frac{2\pi k}{3}\right) + 3i\sin\left(\frac{2\pi k}{3}\right)$

For $k = 0$: $u = 3(\cos 0 + i\sin 0) = 3(1 + 0i) = 3(1) = 3$

For $k = 1$: $u = 3\cos\left(\frac{2\pi}{3}\right) + 3i\sin\left(\frac{2\pi}{3}\right) = -\frac{3}{2} + \frac{3\sqrt{3}}{2}i$

For $k = 2$: $u = 3\cos\left(\frac{4\pi}{3}\right) + 3i\sin\left(\frac{4\pi}{3}\right) = -\frac{3}{2} - \frac{3\sqrt{3}}{2}i$

Logarithms and Exponentials Essential Skills Practice Workbook with Answers

Check these answers by cubing them. The obvious one is $3^3 = 27$. For the others, we will work out one example:

$$\left(-\frac{3}{2}+\frac{3\sqrt{3}}{2}i\right)^3 = \left(-\frac{3}{2}+\frac{3\sqrt{3}}{2}i\right)\left(-\frac{3}{2}+\frac{3\sqrt{3}}{2}i\right)\left(-\frac{3}{2}+\frac{3\sqrt{3}}{2}i\right) = \left(-\frac{3}{2}+\frac{3\sqrt{3}}{2}i\right)\left[\frac{9}{4}-\frac{18\sqrt{3}}{4}i-\frac{9(3)}{4}\right]$$

$$= \left(-\frac{3}{2}+\frac{3\sqrt{3}}{2}i\right)\left(\frac{9}{4}-\frac{9\sqrt{3}}{2}i-\frac{27}{4}\right) = \left(-\frac{3}{2}+\frac{3\sqrt{3}}{2}i\right)\left(-\frac{18}{4}-\frac{9\sqrt{3}}{2}i\right) = \left(-\frac{3}{2}+\frac{3\sqrt{3}}{2}i\right)\left(-\frac{9}{2}-\frac{9\sqrt{3}}{2}i\right)$$

$$= \frac{27}{4}+\frac{27\sqrt{3}}{4}i-\frac{27\sqrt{3}}{4}i+\frac{27(3)}{4} = \frac{27}{4}+\frac{81}{4}+0i = \frac{108}{4} = 27$$

Exercise Set 10.7

1) $a = 2, b = 7, c = 9$.
$$x = \frac{-b\pm\sqrt{b^2-4ac}}{2a} = \frac{-7\pm\sqrt{7^2-4(2)(9)}}{2(2)} = \frac{-7\pm\sqrt{49-72}}{4} = \frac{-7\pm\sqrt{-23}}{4} = \frac{-7\pm i\sqrt{23}}{4}$$

2) $a = 5, b = -10, c = 6$
$$x = \frac{-b\pm\sqrt{b^2-4ac}}{2a} = \frac{-(-10)\pm\sqrt{(-10)^2-4(5)(6)}}{2(5)} = \frac{10\pm\sqrt{100-120}}{10}$$
$$= \frac{10\pm\sqrt{-20}}{10} = \frac{10\pm i\sqrt{20}}{10} = \frac{10\pm 2i\sqrt{5}}{10} = \frac{5\pm i\sqrt{5}}{5} = 1\pm\frac{\sqrt{5}}{5}i \text{ Alternate answers: } 1\pm\frac{i}{\sqrt{5}} \text{ (since } \frac{\sqrt{5}}{5} = \frac{1}{\sqrt{5}}\text{)}$$
Note: $\sqrt{20} = \sqrt{(4)(5)} = \sqrt{4}\sqrt{5} = 2\sqrt{5}$

Note: We divided the numerator and denominator of $\frac{10\pm 2i\sqrt{5}}{10}$ each by 2.

Exercise Set 10.8

1) $\cos i = \cosh 1 = \frac{e^1+e^{-1}}{2} \approx 1.543$

2) $\sin(i\pi) = i\sinh \pi = \frac{e^\pi - e^{-\pi}}{2}i \approx 11.55i$

3) $\cosh(\pi i) = \cos \pi = -1$

4) $\sinh\left(\frac{\pi}{6}i\right) = i\sin\left(\frac{\pi}{6}\right) = \frac{i}{2} = 0.5i$

5) $\tanh\left(\frac{\pi}{4}i\right) = i\tan\left(\frac{\pi}{4}\right) = i(1) = i$

6) $\cosh\left(\frac{5\pi}{6}i\right) = \cos\left(\frac{5\pi}{6}\right) = -\frac{\sqrt{3}}{2} \approx -0.8660$

7) $\sin\left(\frac{\pi}{2}+i\right) = \sin\left(\frac{\pi}{2}\right)\cosh 1 + i\cos\left(\frac{\pi}{2}\right)\sinh 1 = (1)\frac{e^1+e^{-1}}{2}+i(0)\frac{e^1-e^{-1}}{2} \approx 1.543$

Answer Key

8) $\cos\left(\frac{\pi}{6} - i\right) = \cos\left(\frac{\pi}{6}\right)\cosh(-1) - i\sin\left(\frac{\pi}{6}\right)\sinh(-1) = \frac{\sqrt{3}}{2}\frac{e^{-1}+e^{-(-1)}}{2} - i\frac{1}{2}\frac{e^{-1}-e^{-(-1)}}{2}$

$= \frac{\sqrt{3}}{2}\frac{e^{-1}+e^{1}}{2} - i\frac{1}{2}\frac{e^{-1}-e^{1}}{2} \approx 1.336 - (-0.5876i) \approx 1.336 + 0.5876i$

Note: Since $\cosh(-y) = \cosh y$ and $\sinh(-y) = -\sinh y$ (Sec. 6.2, Problems 2-3), we could use the formula $\cos(x - iy) = \cos x \cosh y + i \sin x \sinh y$ to arrive at the same answer.

9) $\sinh(2 + \pi i) = \sinh 2 \cos \pi + i \sin \pi \cosh 2 = \frac{e^{2}-e^{-2}}{2}(-1) + i(0)\frac{e^{2}+e^{-2}}{2} \approx -3.627$

10) $\cosh\left(\frac{1-\pi i}{2}\right) = \cosh\left(\frac{1}{2} - \frac{\pi}{2}i\right) = \cosh\left(\frac{1}{2}\right)\cos\left(-\frac{\pi}{2}\right) + i\sinh\left(\frac{1}{2}\right)\sin\left(-\frac{\pi}{2}\right)$

$= \frac{e^{1/2}+e^{-1/2}}{2}(0) + i\frac{e^{1/2}-e^{-1/2}}{2}(-1) \approx -0.5211i$

Note: An alternative solution can be found using a similar idea to the note for the solution to Problem 8.

WAS THIS BOOK HELPFUL?

A great deal of effort and thought was put into this book, such as:
- Breaking down the solutions to help make the math easier to understand.
- Careful selection of examples and problems for their instructional value.
- Emphasis of what a logarithm means in Chapter 2.
- Coverage of a variety of essential topics and skills.
- Explanations of the ideas behind the math.
- Full solutions to the exercises included in the answer key.

If you appreciate the effort that went into making this book possible, there is a simple way that you could show it:

Please take a moment to post an honest review.

For example, you can review this book at Amazon.com or Goodreads.com.

Even a short review can be helpful and will be much appreciated. If you're not sure what to write, following are a few ideas, though it's best to describe what is important to you.
- How much did you learn from reading and using this workbook?
- Were the solutions at the back of the book helpful?
- Were you able to understand the solutions?
- Was it helpful to follow the examples while solving the problems?
- Would you recommend this book to others? If so, why?

Do you believe that you found a mistake? Please email the author, Chris McMullen, at greekphysics@yahoo.com to ask about it. One of two things will happen:
- You might discover that it wasn't a mistake after all and learn why.
- You might be right, in which case the author will be grateful and future readers will benefit from the correction. Everyone is human.

ABOUT THE AUTHOR

Dr. Chris McMullen has over 20 years of experience teaching university physics in California, Oklahoma, Pennsylvania, and Louisiana. Dr. McMullen is also an author of math and science workbooks. Whether in the classroom or as a writer, Dr. McMullen loves sharing knowledge and the art of motivating and engaging students.

The author earned his Ph.D. in phenomenological high-energy physics (particle physics) from Oklahoma State University in 2002. Originally from California, Chris McMullen earned his Master's degree from California State University, Northridge, where his thesis was in the field of electron spin resonance.

As a physics teacher, Dr. McMullen observed that many students lack fluency in fundamental math skills. In an effort to help students of all ages and levels master basic math skills, he published a series of math workbooks on arithmetic, fractions, long division, algebra, geometry, trigonometry, and calculus entitled *Improve Your Math Fluency*. Dr. McMullen has also published a variety of science books, including astronomy, chemistry, and physics workbooks.

Author, Chris McMullen, Ph.D.

MATH

This series of math workbooks is geared toward practicing essential math skills:
- Prealgebra
- Algebra
- Geometry
- Trigonometry
- Logarithms and exponentials
- Calculus
- Fractions, decimals, and percentages
- Long division
- Multiplication and division
- Addition and subtraction
- Roman numerals
- The four-color theorem and basic graph theory

www.improveyourmathfluency.com

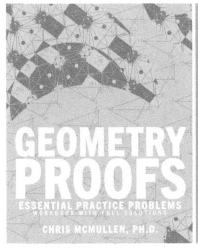
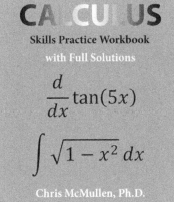
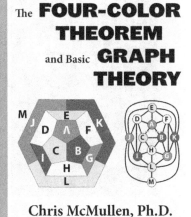

PUZZLES

The author of this book, Chris McMullen, enjoys solving puzzles. His favorite puzzle is Kakuro (kind of like a cross between crossword puzzles and Sudoku). He once taught a three-week summer course on puzzles. If you enjoy mathematical pattern puzzles, you might appreciate:

300+ Mathematical Pattern Puzzles

Number Pattern Recognition & Reasoning
- Pattern recognition
- Visual discrimination
- Analytical skills
- Logic and reasoning
- Analogies
- Mathematics

THE FOURTH DIMENSION

Are you curious about a possible fourth dimension of space?
- Explore the world of hypercubes and hyperspheres.
- Imagine living in a two-dimensional world.
- Try to understand the fourth dimension by analogy.
- Several illustrations help to try to visualize a fourth dimension of space.
- Investigate hypercube patterns.
- What would it be like to be a 4D being living in a 4D world?
- Learn about the physics of a possible four-dimensional universe.

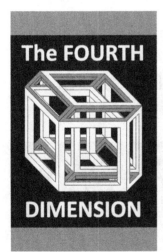

SCIENCE

Dr. McMullen has published a variety of **science** books, including:
- Basic astronomy concepts
- Basic chemistry concepts
- Balancing chemical reactions
- Calculus-based physics textbooks
- Calculus-based physics workbooks
- Calculus-based physics examples
- Trig-based physics workbooks
- Trig-based physics examples
- Creative physics problems
- Modern physics

www.monkeyphysicsblog.wordpress.com

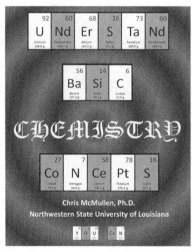
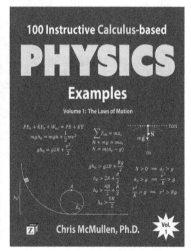

Made in the USA
Las Vegas, NV
10 September 2023